Medical Virology: from Pathogenesis to Disease Control

Series Editor

Shailendra K. Saxena
Centre for Advanced Research, King George's Medical University, Lucknow, India

This book series reviews the recent advancement in the field of medical virology including molecular epidemiology, diagnostics and therapeutic strategies for various viral infections. The individual books in this series provide a comprehensive overview of infectious diseases that are caused by emerging and re-emerging viruses including their mode of infections, immunopathology, diagnosis, treatment, epidemiology, and etiology. It also discusses the clinical recommendations in the management of infectious diseases focusing on the current practices, recent advances in diagnostic approaches and therapeutic strategies. The books also discuss progress and challenges in the development of viral vaccines and discuss the application of viruses in the translational research and human healthcare.

More information about this series at http://www.springer.com/series/16573

Shailendra K. Saxena
Editor

Coronavirus Disease 2019 (COVID-19)

Epidemiology, Pathogenesis, Diagnosis, and Therapeutics

Springer

Editor
Shailendra K. Saxena
Centre for Advanced Research
King George's Medical University
Lucknow, India

ISSN 2662-981X ISSN 2662-9828 (electronic)
Medical Virology: from Pathogenesis to Disease Control
ISBN 978-981-15-4813-0 ISBN 978-981-15-4814-7 (eBook)
https://doi.org/10.1007/978-981-15-4814-7

This Springer imprint is published by the registered company Springer Nature Singapore Pte Ltd.
The registered company address is: 152 Beach Road, #21-01/04 Gateway East, Singapore 189721, Singapore

Dedicated
to my ***Parents*** *and* ***Family***
who believed in academics
as the way forward for an intelligent mind
and
to the ***Teachers***
who introduced me the subject
and nurtured my interest in it.

Foreword

Coronavirus Disease 2019 (COVID-19) caused by the novel coronavirus (SARS-CoV-2) has posed a medical emergency and a global crisis rapidly, from having first emerged in December 2019. On 11th March 2020, it was declared a pandemic by the World Health Organization (WHO). By 6th April 2020, globally, there were more than 1.2 million confirmed cases of COVID-19 and 67,000 deaths across 209 countries, areas, or territories according to WHO updates. Individuals, families, healthcare systems, food systems, and economies are buckling under its strain. Lives have been turned upside down for multitudes and, for some, COVID-19 has caused a humanitarian tragedy.

Coronaviruses (CoVs) represent a large family of viruses, some of which have previously caused severe human diseases such as Middle East respiratory syndrome (MERS) and severe acute respiratory syndrome (SARS). The common symptoms of COVID-19 include fever and cough, and a proportion of patients may develop shortness of breath and other symptoms. In more severe cases, the infection can cause pneumonia, acute respiratory distress syndrome, organ failure, and death. The improved management of COVID-19 human infections will depend on characterization of the SARS-CoV-2 virus, including its transmission capability, severity of resulting infection, and availability of vaccines and antiviral drugs to control the impact of COVID-19. Public health measures are clearly important to reduce its transmission: thorough and frequent handwashing, hygiene, social distancing, and self-isolation as indicated.

The international medical, scientific, and public health communities and government agencies have engaged urgently with the challenge, and several reliable websites, scientific publications, and other resources have followed. Concomitantly, there has been a lot of misinformation and myths that have spread like wildfire through social media and other channels, which cause confusion and are frankly harmful. The Director General of the WHO warned that "We're not just fighting an epidemic; we're fighting an infodemic."

It is vital to engage responsibly with all aspects of the COVID-19 emergency. The present book *Coronavirus Disease 2019 (COVID-19): Epidemiology, Pathogenesis,*

Diagnosis, and Therapeutics from Springer Nature focuses on several aspects of SARS-CoV2 and COVID-19. This book includes information on current insights into COVID-19, focusing on epidemiology, structure of SARS-CoV-2, genome organization, replication and pathogenesis, transmission cycle, and host immune response during SARS-CoV-2 infection. The chapters cover information on a range of topics from clinical characteristics and differential clinical diagnosis of COVID-19, laboratory diagnosis and development of therapeutics as well as preventive strategies, SARS-CoV-2 infection among children and adolescents, and other relevant issues.

The book will be of interest to a wide range of professionals, placing a comprehensive set of information in a single collection. Professor Shailendra Saxena, his team, and chapter contributors are to be congratulated for their timely effort in conceptualizing and swiftly summarizing the current knowledge in this book. It is important, too, to note the dynamic nature of this global threat and responses to it across different countries and globally. The current knowledge and understanding about COVID-19 are constantly unfolding, being updated and expanded through ongoing research efforts as well as learning by experience on a daily basis. Therefore, resources from the World Health Organization and country-specific guidance and medical updates should also be used appropriately.

This is a time when professionals and agencies from all potentially relevant disciplines need to work in a concerted way, including those working in and outside the fields of virology and infectious diseases. It is in this spirit as an epidemiologist and public health physician that I applaud this current effort. I hope this book together with dynamically updated and other reliable sources will stimulate different options to formulate and build ideas and galvanize actions to combat COVID-19 at speed.

University of Cambridge
School of Clinical Medicine
Medical Research Council (MRC)
Epidemiology Unit
Cambridge, United Kingdom

Nita G. Forouhi
Programme Leader and Consultant
Public Health Physician

Preface

Novel coronavirus (SARS-CoV-2) causing coronavirus disease (COVID-19) has become a pandemic and led to at least 117,021 deaths as of 14th April 2020 globally. The transmission of SARS-CoV-2 occurs via close contact with the respiratory droplets generated by infected individuals. COVID-19 patients have been shown to undergo acute respiratory distress syndrome, which is defined as cytokine storm and is the foremost reason for morbidity and mortality due to multiple organ failure. Coronaviruses are zoonotic in nature, which means that they originate from animal sources and may jump to human population. Coronaviruses (CoVs) are a large family of viruses that cause illnesses ranging from the common cold to more severe diseases such as Middle East respiratory syndrome (MERS-CoV) and severe acute respiratory syndrome (SARS-CoV). SARS-CoV-2 is a novel coronavirus that has not been previously identified in humans. Even with the implementation of strong travel restrictions, international travel by a large number of individuals exposed to SARS-CoV-2 has resulted in the spread of the virus worldwide.

At the global level, scientific communities including various healthcare professionals are working continuously to understand the virological, epidemiological, and clinical aspect of COVID-19 in order to identify effective drugs and vaccines. In this regard, this book novel *Coronavirus Disease 2019 (COVID-19)* provides a comprehensive understanding of COVID-19 with its associated challenges for its treatment and prevention to protect the human race. The book focuses on crucial virological aspects of SARS-CoV-2 including its epidemiology, structure, genome organization, replication, pathogenesis, host immune response, and transmission cycle. The various aspects of the book cover the clinical characteristics, differential clinical diagnosis of COVID-19 patients in children and adolescents, diagnosis, development of therapeutics as well as preventive strategies. In addition, the book also emphasizes on the emergence and re-emergence of SARS-CoV, classical coronavirus infection, and prepares the world for the perpetual challenges of future pandemics. I hope this book will provide a readily available resource in this area and may help multidisciplinary teams to focus on the treatment and management of the disease effectively. As a result, this book focuses on the understanding of COVID-19 from various perspectives while discussing the crucial aspect of the disease and its associated pathobiology. The book will be of interest to scientists, clinicians, healthcare professionals, research and development and policy-makers, and may contribute towards the eradication of SARS-CoV-2.

This book provides a comprehensive overview of COVID-19 and plausible strategies and schemes for their mitigation at national and international levels. However, due to rapidly changing information and guidelines, kindly refer to the World Health Organization (WHO) and country-specific guidelines appropriately. The information produced in this book is not intended for direct diagnostic use or medical decision-making without review and oversight by a clinical professional. Individuals should not change their health behavior solely on the basis of information produced in this book. The publisher/editors/contributors do not independently verify the validity or utility of the information produced in this book. If you have questions about the information produced in this book, please see a healthcare professional. All these aspects of the book are imperative for safeguarding the human race from more loss of resources and economies due to COVID-19. I hope this work might increase the interest in this field of research and that the reader will find it useful for their investigations, management, and clinical usage.

Lucknow, India Shailendra K. Saxena

Suggested Reading for Updated Information About COVID-19

- World Health Organization. Coronavirus disease (COVID-19) pandemic. https://www.who.int/emergencies/diseases/novel-coronavirus-2019
- Centers for Disease Control and Prevention. United States Department of Health and Human Services. Coronavirus (COVID-19). https://www.cdc.gov/coronavirus/2019-ncov/index.html
- Food and Drug Administration. United States Department of Health and Human Services. Coronavirus Disease 2019 (COVID-19). https://www.fda.gov/emergency-preparedness-and-response/counterterrorism-and-emerging-threats/coronavirus-disease-2019-covid-19
- European Centre for Disease Prevention and Control. An agency of the European Union. COVID-19. https://www.ecdc.europa.eu/en/covid-19-pandemic
- Ministry of Health and Family Welfare, Government of India. COVID-19 India. https://www.mohfw.gov.in/
- Indian Council of Medical Research, Government of India. COVID-19. https://icmr.nic.in/content/covid-19
- MyGov Cell. National Informatics Centre, Ministry of Electronics & Information Technology, Government of India. IndiaFightsCorona COVID-19. https://mygov.in/covid-19/

Acknowledgments

This book was conceptualized considering the unprecedented pandemic of COVID-19 and to focus on various aspects of novel coronavirus infections such as global epidemiology, genome organization, immunopathogenesis, transmission cycle, diagnosis, treatment, prevention, and control strategies including mental health and pandemic preparation. All these aspects are imperative for safeguarding the human race from more loss of resources and economies due to COVID-19. To overcome these issues and fill the gap, we hope this book shall provide a readily available resource in this area. The aim of this book is to acknowledge the potential of COVID-19 and remedies to mitigate it.

I am overwhelmed in all humbleness and gratefulness to acknowledge my in-depth gratitude to all the contributors who trusted me and supported this work. I hope they are as proud of this book as I am. I would also like to thank Springer Nature Publisher and all team members to not only consider this book for publication but also publish in the shortest possible time, despite COVID-19 pandemic-associated international restrictions, lockdown, work from home, and several other limiting factors. All the reports referred to in this book are given proper citation. However, any missed information is just unintentional and explicable.

My research fellows and students are central to all my research and academic work. They are the motivating force behind anything constructive I do. They are truly brilliant and have a bright future. I would like to express my special thanks of gratitude to my mentors, teachers, and students who gave me the strength to accomplish this. Also, I would like to thank my colleagues, family, and friends who provided a lot of encouragement and support during the work on this book.

A happy environment at home is essential for any kind of growth, and I thank my family, especially my talented wife and children, for the same.

Lucknow, India — Shailendra K. Saxena

ACKNOWLEDGMENTS

[illegible]

About This Book

SARS-CoV-2 is a new virus responsible for an outbreak of respiratory illness known as novel coronavirus disease 2019 (COVID-19), which has spread to several countries around the world. Although we are in the twenty-first century and have advanced technologies in hand, COVID-19 is responsible for enormous mortality due to scarcity of effective differential diagnostics, therapeutics, and preventive strategies. This book provides a comprehensive overview of recent trends in COVID-19 and novel coronavirus (SARS-CoV-2) infections, its biology, and associated challenges for the treatment and prevention. This book focuses on various aspects of novel coronavirus infections such as global epidemiology, genome organization, immunopathogenesis, transmission cycle, diagnosis, treatment, prevention, and control strategies including mental health and pandemic preparation. All these aspects of the book are imperative for safeguarding the human race from more loss of resources and economies due to global SARS-CoV-2 infections. To overcome these issues and fill the gap, I introduce this book to provide a readily available resource in this area. However, for the most recent updates and guidelines, kindly refer to the World Health Organization (WHO) website and/or relevant country guidelines. As a leading research publisher, Springer Nature is committed to supporting the global response to emerging outbreaks by enabling fast and direct access to the latest available research, evidence, and data. This book shall provide recent advancements in this pandemic disease COVID-19.

About This Book

Contents

About the Editor

Shailendra K. Saxena is Vice Dean and Professor at King George's Medical University, Lucknow. His primary research interest is to understand molecular mechanisms of host defense during human viral infections and to develop new strategies for predictive, preventive, and therapeutic strategies for them. Dr. Saxena has received specialized training in COVID-19: tackling the novel coronavirus by the London School of Hygiene & Tropical Medicine and UK Public Health Rapid Support Team; and in emerging respiratory viruses, including COVID-19: methods for detection, prevention, response, and control; infection prevention and control (IPC) for novel coronavirus (COVID-19); and COVID-19: Operational Planning Guidelines and COVID-19 Partners Platform to support country preparedness and response by The United Nations Country Team (UNCT)-World Health Organization's Health Emergencies Programme (WHE). Prof. Saxena's work has been published in reputed international journals with high impact factor. His work has been highly cited by numerous investigators globally and honored by several prestigious national and international awards, fellowships, and scholarships in India and abroad, including various Young Scientist Awards and BBSRC India Partnering Award. In addition, he was named the Global Leader in Science by *The Scientist* magazine (USA) and International Opinion Leader/Expert involved in the vaccination for JE by IPIC (UK). Prof. Saxena has been elected Fellow of The Royal Society of Biology, UK (FRSB); The Royal Society of Chemistry, UK (FRSC); The Academy of

Environmental Biology, India (FAEB); Indian Virological Society (FIVS); and The Biotech Research Society, India (FBRS) and received a Fellowship of the (European) Academy of Translational Medicine Professionals, Austria (FacadTM). He has been awarded Dr. JC Bose National Award by the Department of Biotechnology (DBT, Ministry of Science and Technology, Government of India) in Biotechnology and has active collaboration with US universities.

Chapter 1
Current Insight into the Novel Coronavirus Disease 2019 (COVID-19)

Shailendra K. Saxena, Swatantra Kumar, Vimal K. Maurya, Raman Sharma, Himanshu R. Dandu, and Madan L. B. Bhatt

Abstract SARS-CoV-2 is a novel strain of coronavirus that has not been previously identified in humans. It has been declared a pandemic and has infected at least 1,844,683 individuals and caused 117,021 deaths as of 14th April 2020. Transmission among humans occurs via close contact with an infected individual that produces respiratory droplets. Patients have been shown to undergo acute respiratory distress syndrome, which is defined as cytokine storm. The diagnosis relies on detection of nucleic acid, IgG/IgM antibodies, and a chest radiograph of the suspected individuals. The genome of SARS-CoV-2 is similar to other coronaviruses that comprise of ten open reading frames (ORFs). SARS-CoV-2 spike protein exhibits higher affinity to ACE2 receptor as compared with SARS-CoV. Repurposing drugs like favipiravir, remdesivir, chloroquine, and TMPRSS2 protease inhibitors have been shown to be effective for the treatment of COVID-19. Personal protective measures should be followed to prevent SARS-CoV-2 infection. In addition, a clinical trial of SARS-CoV-2 vaccine, mRNA-1273, has been started. This chapter provides a glimpse of advancements made in the area of SARS-CoV-2 infection by proving recent clinical and research trials in the field.

S. K. Saxena (✉)
Centre for Advanced Research (CFAR)-Stem Cell/Cell Culture Unit, Faculty of Medicine, King George's Medical University (KGMU), Lucknow, India

Department of Medicine, Sawai Man Singh Medical College, Jaipur, India
e-mail: shailen@kgmcindia.edu

S. Kumar · V. K. Maurya · M. L. B. Bhatt
Centre for Advanced Research (CFAR), Faculty of Medicine, King George's Medical University (KGMU), Lucknow, India

R. Sharma
Department of Medicine, Sawai Man Singh Medical College, Jaipur, India

H. R. Dandu
Department of Internal Medicine, King George's Medical University, Lucknow, India

S. K. Saxena (ed.), *Coronavirus Disease 2019 (COVID-19)*, Medical Virology: from Pathogenesis to Disease Control, https://doi.org/10.1007/978-981-15-4814-7_1

Keywords SARS-CoV-2 · COVID-19 · Novel coronavirus · Cytokine storm · Favipiravir · Remdesivir · Chloroquine · mRNA-1273

1.1 Introduction

On 11 March 2020, the World Health Organization (WHO) declared severe acute respiratory syndrome coronavirus 2 (SARS-CoV-2) as a pandemic that causes novel coronavirus disease 2019 (COVID-19) (World Health Organization 2020a). By 14 April 2020, around 1,844,683 confirmed cases with 117,021 deaths were reported from at least 213 countries, areas, or territories (World Health Organization 2020b). SARS-CoV-2 is a novel strain of coronavirus that has not been previously identified in humans. Phylogenetic analysis suggests that SARS-CoV-2 might have emerged from the zoonotic cycle and rapidly spread by human to human transmission (Chan et al. 2020a). However, the exact source of SARS-CoV-2 has not been identified yet. Transmission among humans occurs via close contact with an infected individual that produces respiratory droplets while coughing or sneezing within a range of about 6 ft (Ghinai et al. 2020). Infected individuals have been reported with common clinical symptoms involving fever, nonproductive cough, myalgia, shortness of breath, as well as normal or decreased leukocyte counts (Fig. 1.1) (Zhang et al. 2020). In addition, severe cases of infection cause pneumonia, severe acute respiratory syndrome, kidney failure, and death (Zhao et al. 2020; Xiong et al. 2020). Even with the implementation of strong travel restrictions, a large number of individuals exposed to SARS-CoV-2 have been traveling internationally without being detected, leading to spread of the virus worldwide (Chinazzi et al. 2020). However, extensive measures have been implemented by outstanding public health action to reduce person-to-person transmission of SARS-CoV-2 (Fig. 1.2). In addition, the scientific fraternity worldwide has been continuously working on COVID-19 from the beginning by publishing the genome and developing highly specific diagnostic tools for the detection of SARS-CoV-2 infection.

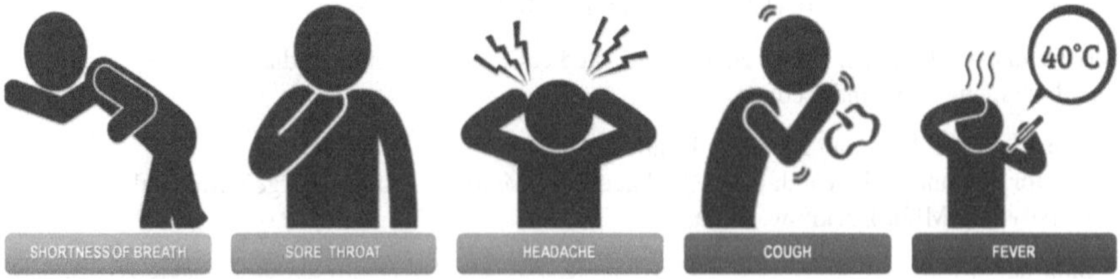

Fig. 1.1 Typical symptoms of COVID-19

Fig. 1.2 Personal protective measures to prevent SARS-CoV-2 infection

1.2 SARS-CoV-2 Genome and Pathogenesis

SARS-CoV-2 is a single-stranded RNA virus of ~30 kb genome size, which belongs to the genus *Coronavirus* and family Coronaviridae. The genome of SARS-CoV-2 is similar to other coronaviruses that comprise of ten open reading frames (ORFs). The first ORFs (ORF1a/b), about two-thirds of viral RNA, are translated into two large polyproteins pp1a and pp1ab, which processed into non-structural proteins (nsp1-nsp16) (Chan et al. 2020b). The size of each SARS-CoV-2 virion is about 70–90 nm (Kim et al. 2020). The genome of SARS-CoV-2 encodes for four structural proteins similar to other coronaviruses. These proteins are S (spike), E (envelope), M (membrane), and N (nucleocapsid) protein which are required to make complete virus particle. S protein is responsible for the attachment and entry of SARS-CoV-2 to the host target cell receptor, probably angiotensin-converting enzyme 2 (ACE2) mainly expressed on alveolar epithelial type II (AECII) cells, including extrapulmonary tissues such as heart, kidney, endothelium, and intestine (Yan et al. 2020). SARS-CoV-2 has been shown to exhibit novel glycosylation sites in the spike glycoprotein of 2019-nCoV, suggesting that the virus may utilize different glycosylation sites to interact with its receptors (Kumar et al. 2020). Studies have demonstrated that SARS-CoV-2 spike protein has higher affinity to the ACE2 receptor as compared with SARS (Walls et al. 2020).

1.3 Host Immune Response Against SARS-CoV-2

Upon entry into the host target cells, the viral antigens get presented via antigen-presenting cells (APCs) to virus-specific cytotoxic T lymphocytes (CTL). So far, studies have not been conducted that reveal the peptide presentation. However, CTL epitopes of SARS-CoV-2 have been predicted by several studies, which may be used for understanding the pathogenesis and development of peptide-based vaccines (Kumar et al. 2020; Walls et al. 2020). Studies have been conducted in SARS-CoV-2 infected patients showing the activation and reduction in $CD4^+$ and $CD8^+$ T cell counts (Li et al. 2020a). In addition, SARS-CoV-2 patients have been found to present with acute respiratory distress syndrome (ARDS) (Zumla et al. 2020). ARDS is a cytokine storm syndrome (CSS) which is a lethal uncontrollable inflammatory response resulting from the release of large pro-inflammatory cytokines (IL-1β, IFN-α, IFN-γ, IL-12, IL-6, IL-18, TNF-α, IL-33, TGFβ, etc.) and chemokines (CCL3, CCL2, CXCL8, CCL5, CXCL9, CXCL10, etc.) by immune cells (Li et al. 2020a).

1.4 Diagnosis of Human SARS-CoV-2 Infection

Suspected patients get diagnosed for SARS-CoV-2 infection by collecting various specimens, including nasopharyngeal or oropharyngeal swabs, nasopharyngeal or oropharyngeal aspirates or washes, bronchoalveolar lavage, sputum, tracheal aspirates, and blood. Specimens can be stored at 4 °C for up to 72 h after sample collection and may be stored at −70 °C for longer periods of time (Centre for Disease Control and Prevention 2020a). Diagnosis tests such as nucleic acid test, ELISA, CT scan, and blood cultures are being implemented for the detection of SARS-CoV-2 infection. Commonly used nucleic acid tests are RT-qPCR and high-throughput sequencing, where RT-qPCR is the effective and straightforward method for detection of pathogenic viruses in respiratory secretions and blood. Specific primers and probes against ORF1ab and N gene regions have been recommended to use for the detection of SARS-CoV-2 (Wang et al. 2020a). In addition, immunological detection of IgM and IgG antibodies are being performed to diagnose the COVID-19 patients (Li et al. 2020b). Patients reporting respiratory discomfort were evaluated using CT scan (Zhou et al. 2020).

1.5 Treatment and Drugs for SARS-CoV-2

There is no specific treatment available for SARS-CoV-2 and the current treatment relies on supportive care of the infected patients (Centre for Disease Control and Prevention 2020b). However, some evidences suggest the use of repurposing drugs as the current choice of therapy. Remdesivir, a nucleoside analogue-based drug that is currently under clinical trial for treating Ebola virus infection, has been shown to block SARS-CoV-2 infection in vitro (Wang et al. 2020b). In addition, favipiravir, a type of RNA-dependent RNA polymerase inhibitor that has been designed to treat influenza virus infection, has been found to exhibit antiviral activity against SARS-CoV-2 (Dong et al. 2020). Use of chloroquine, especially hydroxychloroquine, has been found to be effective against SARS-CoV-2 in vitro, which interferes with the glycosylation of cellular receptors (Yao et al. 2020). Apart from attachment inhibitors, TMPRSS2 protease inhibitors have also been found to block SARS-CoV-2 infection in lung cells (Hoffmann et al. 2020).

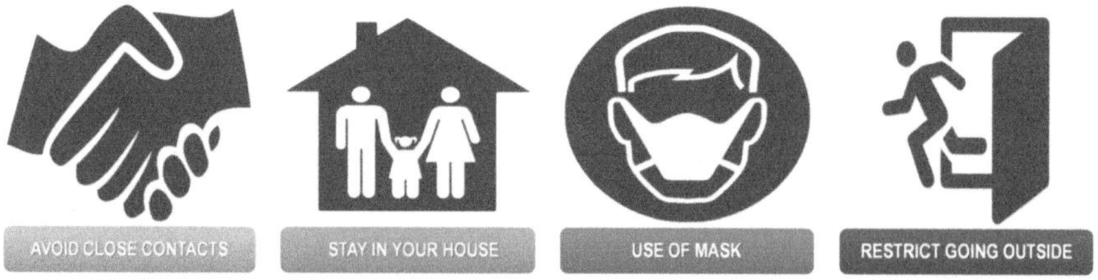

Fig. 1.3 Steps needed to be taken by COVID-19 patients in order to prevent the spread of SARS-CoV-2 infection

1.6 Prevention and Control Strategies for SARS-CoV-2 Infection

The current preventive strategies of SARS-CoV-2 infection relies on personal protective measures such as covering of nose/mouth when coughing or sneezing, use of FFP3 or N95 mask, use of tissues to contain respiratory secretions and dispose of these in nearest waste receptacle, and hand hygiene after contact with contaminated objects/materials or respiratory secretion (Fig. 1.3) (Centre for Disease Control and Prevention 2020c). Healthcare professionals are at the highest risk of getting SARS-CoV-2 infection from infected patients and therefore extreme precaution needs to be taken while handling COVID-19 patients. International travelers presenting any symptoms of SARS-CoV-2 should be isolated and quarantined to prevent further infections (Gostic et al. 2020). Apart from these personal protective measures, development of effective vaccine is the ultimate way of controlling SARS-CoV-2 infection. Using bioinformatics approaches, novel cytotoxic T lymphocyte (CTL) generated from spike glycoproteins may be used to develop effective vaccine for SARS-CoV-2. The first SARS-CoV-2 vaccine, namely, mRNA-1273, is under clinical trial and involves 45 volunteers who have received two intramuscular injections at an interval of 28 days (U.S. National Library of Medicine 2020).

1.7 Conclusions

SARS-CoV-2 has been declared a pandemic that causes COVID-19. Infected individuals have been reported with common flu-like symptoms and cytokine storm syndrome in severe cases. Diagnosis of COVID-19 relies on the detection of nucleic acid tests by RT-qPCR. Current treatment relies on the symptomatic relief of the patients. However, several repurposing drugs like favipiravir, remdesivir, chloroquine, and TMPRSS2 protease inhibitors have been shown to be effective. Personal protective measures should be followed to prevent SARS-CoV-2 infection. In addition, a clinical trial of SARS-CoV-2 vaccine, mRNA-1273, has been started.

1.8 Future Perspectives

The emergence of any infectious viral disease is difficult to anticipate, even though use of spatial epidemiology and mathematical modeling may predict the occurrence of emerging or re-emerging diseases like COVID-19. RNA recombination, mutation, and reassortment as well as other factors including globalization, expanding human population, deforestation, and altered ecosystems are the most convergent forces for the emergence of viral infectious diseases. SARS-CoV-2 infection is of global public health and economic importance and therefore needs collective government and societal response. Considering the escalating number of cases worldwide, the WHO has declared SARS-CoV-2 as a pandemic. The global shutdown of trade and travel may result in a reduction in the transmission rate of SARS-CoV-2. Personal protective measures should be implemented to reduce the risk of SARS-CoV-2. In order to prevent the spread of SARS-CoV-2 infection, several major steps should be taken, for instance strengthening surveillance and conducting awareness programs. In addition, extensive applied research should be funded and executed to understand the molecular mechanism and to develop effective prevention and control strategies for COVID-19.

Acknowledgments We are grateful to the Vice Chancellor, King George's Medical University (KGMU), Lucknow, India, for the encouragement of this work. The authors have no other relevant affiliations or financial involvement with any organization or entity with a financial interest in or financial conflict with the subject matter or materials discussed in the manuscript apart from those disclosed.

References

Centre for Disease Control and Prevention (2020a) Interim guidelines for collecting, handling, and testing clinical specimens from persons for coronavirus disease 2019 (COVID-19). Centre for Disease Control and Prevention. https://www.cdc.gov/coronavirus/2019-ncov/lab/guidelines-clinical-specimens.html. Accessed 18 Mar 2020

Centre for Disease Control and Prevention (2020b) Interim clinical guidance for management of patients with confirmed coronavirus disease (COVID-19). Centre for Disease Control and Prevention. https://www.cdc.gov/coronavirus/2019-ncov/hcp/clinical-guidance-management-patients.html. Accessed 18 Mar 2020

Centre for Disease Control and Prevention (2020c) Interim infection prevention and control recommendations for patients with suspected or confirmed coronavirus disease 2019 (COVID-19) in healthcare settings. Centre for Disease Control and Prevention. https://www.cdc.gov/coronavirus/2019-ncov/infection-control/control-recommendations.html. Accessed 18 Mar 2020

Chan JF, Yuan S, Kok KH, To KK, Chu H, Yang J, Xing F, Liu J, Yip CC, Poon RW, Tsoi HW, Lo SK, Chan KH, Poon VK, Chan WM, Ip JD, Cai JP, Cheng VC, Chen H, Hui CK, Yuen KY (2020a) A familial cluster of pneumonia associated with the 2019 novel coronavirus indicating person-to-person transmission: a study of a family cluster. Lancet 395(10223):514–523. https://doi.org/10.1016/S0140-6736(20)30154-9

Chan JF, Kok KH, Zhu Z, Chu H, To KK, Yuan S, Yuen KY (2020b) Genomic characterization of the 2019 novel human-pathogenic coronavirus isolated from a patient with atypical pneumonia after visiting Wuhan. Emerg Microbes Infect 9(1):221–236. https://doi.org/10.1080/22221751.2020.1719902

Chinazzi M, Davis JT, Ajelli M, Gioannini C, Litvinova M, Merler S, Pastore Y Piontti A, Mu K, Rossi L, Sun K, Viboud C, Xiong X, Yu H, Halloran ME, Longini IM Jr, Vespignani A (2020) The effect of travel restrictions on the spread of the 2019 novel coronavirus (COVID-19) outbreak. Science. pii: eaba9757. https://doi.org/10.1126/science.aba9757

Dong L, Hu S, Gao J (2020) Discovering drugs to treat coronavirus disease 2019 (COVID-19). Drug Discov Ther 14(1):58–60. https://doi.org/10.5582/ddt.2020.01012

Ghinai I, McPherson TD, Hunter JC, Kirking HL, Christiansen D, Joshi K, Rubin R, Morales-Estrada S, Black SR, Pacilli M, Fricchione MJ, Chugh RK, Walblay KA, Ahmed NS, Stoecker WC, Hasan NF, Burdsall DP, Reese HE, Wallace M, Wang C, Moeller D, Korpics J, Novosad SA, Benowitz I, Jacobs MW, Dasari VS, Patel MT, Kauerauf J, Charles EM, Ezike NO, Chu V, Midgley CM, Rolfes MA, Gerber SI, Lu X, Lindstrom S, Verani JR, Layden JE; Illinois COVID-19 Investigation Team (2020) First known person-to-person transmission of severe acute respiratory syndrome coronavirus 2 (SARS-CoV-2) in the USA. Lancet. pii: S0140-6736(20)30607-3. https://doi.org/10.1016/S0140-6736(20)30607-3

Gostic K, Gomez AC, Mummah RO, Kucharski AJ, Lloyd-Smith JO (2020) Estimated effectiveness of symptom and risk screening to prevent the spread of COVID-19. Elife. 9. pii: e55570. https://doi.org/10.7554/eLife.55570

Hoffmann M, Kleine-Weber H, Schroeder S, Krüger N, Herrler T, Erichsen S, Schiergens TS, Herrler G, Wu NH, Nitsche A, Müller MA, Drosten C, Pöhlmann S (2020) SARS-CoV-2 cell entry depends on ACE2 and TMPRSS2 and is blocked by a clinically proven protease inhibitor. Cell. pii: S0092-8674(20)30229-4. https://doi.org/10.1016/j.cell.2020.02.052

Kim JM, Chung YS, Jo HJ, Lee NJ, Kim MS, Woo SH, Park S, Kim JW, Kim HM, Han MG (2020) Identification of coronavirus isolated from a patient in Korea with COVID-19. Osong Public Health Res Perspect 11(1):3–7. https://doi.org/10.24171/j.phrp.2020.11.1.02

Kumar S, Maurya VK, Prasad AK et al (2020) Structural, glycosylation and antigenic variation between 2019 novel coronavirus (2019-nCoV) and SARS coronavirus (SARS-CoV). VirusDis 31(1):13–21. https://doi.org/10.1007/s13337-020-00571-5

Li X, Geng M, Peng Y, Meng L, Lu S (2020a) Molecular immune pathogenesis and diagnosis of COVID-19. J Pharmaceut Anal

Li Z, Yi Y, Luo X, Xiong N, Liu Y, Li S, Sun R, Wang Y, Hu B, Chen W, Zhang Y, Wang J, Huang B, Lin Y, Yang J, Cai W, Wang X, Cheng J, Chen Z, Sun K, Pan W, Zhan Z, Chen L, Ye F (2020b) Development and clinical application of a rapid IgM-IgG combined antibody test for SARS-CoV-2 infection diagnosis. J Med Virol. https://doi.org/10.1002/jmv.25727

U.S. National Library of Medicine (2020) Safety and immunogenicity study of 2019-nCoV vaccine (mRNA-1273) to prevent SARS-CoV-2 infection. NCT04283461. https://clinicaltrials.gov/ct2/show/NCT04283461

Walls AC, Park YJ, Tortorici MA, Wall A, McGuire AT, Veesler D (2020) Structure, function, and antigenicity of the SARS-CoV-2 spike glycoprotein. Cell. pii: S0092-8674(20)30262-2. https://doi.org/10.1016/j.cell.2020.02.058

Wang Y, Kang H, Liu X, Tong Z (2020a) Combination of RT-qPCR testing and clinical features for diagnosis of COVID-19 facilitates management of SARS-CoV-2 outbreak. J Med Virol. https://doi.org/10.1002/jmv.25721. [Epub ahead of print]

Wang M, Cao R, Zhang L, Yang X, Liu J, Xu M, Shi Z, Hu Z, Zhong W, Xiao G (2020b) Remdesivir and chloroquine effectively inhibit the recently emerged novel coronavirus (2019-nCoV) in vitro. Cell Res 30(3):269–271. https://doi.org/10.1038/s41422-020-0282-0

World Health Organization (2020a) Coronavirus disease 2019 (COVID-19) situation report—51. World Health Organization. https://www.who.int/docs/default-source/coronaviruse/situation-reports/20200311-sitrep-51-covid-19.pdf?sfvrsn=1ba62e57_10. Accessed 16 Mar 2020

World Health Organization (2020b) Coronavirus disease 2019 (COVID-19) situation report—85. World Health Organization. https://www.who.int/docs/default-source/coronaviruse/situation-reports/20200414-sitrep-85-covid-19.pdf?sfvrsn=7b8629bb_4. Accessed 14 Apr 2020

Xiong Y, Sun D, Liu Y, Fan Y, Zhao L, Li X, Zhu W (2020) Clinical and high-resolution CT features of the COVID-19 infection: comparison of the initial and follow-up changes. Investig Radiol. https://doi.org/10.1097/RLI.0000000000000674

Yan R, Zhang Y, Li Y, Xia L, Guo Y, Zhou Q (2020) Structural basis for the recognition of the SARS-CoV-2 by full-length human ACE2. Science. pii: eabb2762. https://doi.org/10.1126/science.abb2762

Yao X, Ye F, Zhang M, Cui C, Huang B, Niu P, Liu X, Zhao L, Dong E, Song C, Zhan S, Lu R, Li H, Tan W, Liu D (2020) In vitro antiviral activity and projection of optimized dosing design of hydroxychloroquine for the treatment of severe acute respiratory syndrome coronavirus 2 (SARS-CoV-2). Clin Infect Dis. pii: ciaa237. https://doi.org/10.1093/cid/ciaa237. [Epub ahead of print]

Zhang JJ, Dong X, Cao YY, Yuan YD, Yang YB, Yan YQ, Akdis CA, Gao YD (2020) Clinical characteristics of 140 patients infected with SARS-CoV-2 in Wuhan, China. Allergy. https://doi.org/10.1111/all.14238. [Epub ahead of print]

Zhao D, Yao F, Wang L, Zheng L, Gao Y, Ye J, Guo F, Zhao H, Gao R (2020) A comparative study on the clinical features of COVID-19 pneumonia to other pneumonias. Clin Infect Dis. pii: ciaa247. https://doi.org/10.1093/cid/ciaa247

Zhou S, Wang Y, Zhu T, Xia L (2020) CT features of coronavirus disease 2019 (COVID-19) pneumonia in 62 patients in Wuhan, China. AJR Am J Roentgenol 1–8. https://doi.org/10.2214/AJR.20.22975

Zumla A, Hui DS, Azhar EI, Memish ZA, Maeurer M (2020) Reducing mortality from 2019-nCoV: host-directed therapies should be an option. Lancet 395(10224):e35–e36. https://doi.org/10.1016/S0140-6736(20)30305-6

Chapter 2
Global Trends in Epidemiology of Coronavirus Disease 2019 (COVID-19)

Nishant Srivastava, Preeti Baxi, R. K. Ratho, and Shailendra K. Saxena

Abstract In December 2019, suddenly 54 cases of viral pneumonia emerged in Wuhan, China, caused by some unknown microorganism. The virus responsible for these pneumonia infections was identified as novel coronavirus of the family Coronaviridae. The novel coronavirus was renamed as COVID-19 by WHO. Infection from the virus has since increased exponentially and has spread all over the world in more than 196 countries. The WHO has declared a Public Health Emergency of International Concern due to the outbreak of COVID-19. The virus is highly infectious and can cause human-to-human transmission. Every 24 h, cases of COVID-19 increase severalfolds. The WHO is monitoring the SARS-CoV-2 spread very closely via a global surveillance system. The current situation demands the enforcement of strict laws which would help in inhibiting the further spread of COVID-19. Social distancing, international travel restrictions to affected countries, and hygiene are three important ways to nullify SARS-CoV-2.Government and private organizations need to come forward and work together during this pandemic. Public awareness, social distancing, and sterilization must be maintained to neutralize the viral infection, especially in major hot spots.

Nishant Srivastava and Preeti Baxi contributed equally as first author.

N. Srivastava
Department of Biotechnology, Meerut Institute of Engineering and Technology, Meerut, India

P. Baxi
Phytosanitary Laboratory, Department of Plant Molecular Biology and Biotechnology, Indira Gandhi Agriculture University, Raipur, India

R. K. Ratho
Department of Virology, Post Graduate Institute of Medical Education and Research, Chandigarh, India

S. K. Saxena (✉)
Centre for Advanced Research (CFAR)-Stem Cell/Cell Culture Unit, Faculty of Medicine, King George's Medical University (KGMU), Lucknow, India
e-mail: shailen@kgmcindia.edu

S. K. Saxena (ed.), *Coronavirus Disease 2019 (COVID-19)*, Medical Virology: from Pathogenesis to Disease Control, https://doi.org/10.1007/978-981-15-4814-7_2

Keywords Coronavirus · COVID-19 · Epidemiology · SARS-CoV-2 · Global trends

Abbreviations

CDC	Central Drug Council
CoV	Coronavirus
COVID	Coronavirus disease
HIV	Human immunodeficiency virus
MERS	Middle East respiratory syndrome
NIH	National Institutes of Health
SARS	Severe Acute Respiratory Syndrome
USA	United States of America
WHO	World Health Organization

2.1 Introduction

Infectious diseases have emerged as major threats to human existence since centuries and can devastate entire populations. Epidemiological studies have shown that millions of lives vanished due to these pandemic outbreaks. Epidemiology is defined by WHO as follows: "Epidemiology is the study of the distribution and determinants of health-related states or events (including disease), and the application of this study to the control of diseases and other health problems. Various methods can be used to carry out epidemiological investigations: surveillance and descriptive studies can be used to study distribution; analytical studies are used to study determinants."

In the past 100 years, the human race has encountered several epidemic diseases mostly associated with viruses. The twentieth century started with the outbreak of the pandemic H1N1 influenza virus in 1918, infecting one-third of the world's population and accounting for 50 million lives worldwide. It is known as the most deadly pandemic in the history of mankind. The influenza virus struck again in 1957–1958 in the form of the H2N2 influenza A virus, triggering a pandemic that claimed 1.1 million lives worldwide. Pandemic outbreaks are not new and occur from time to time. The major problem is controlling and developing effective solutions for these outbreaks as well as monitoring the viruses and other microbes closely for their mutations and cross-genetic translation. Table 2.1 provides a summary of major pandemic outbreaks that have occurred in the history of mankind till date (as per available sources; WHO, CDC, seeker.com and mphonline.com) (Lamb 2013; Bai et al. 2020; Staff 2020).

The outbreak of SARS-CoV-2 was first reported at Wuhan, China, in late December 2019. Initially the infection emerged as viral pneumonia from unknown microbial agents (Lu et al. 2020). The Chinese Center for Disease Control and Prevention identified the virus as novel coronavirus from the throat swab sample of an infected patient on January 7, 2020 (Chen et al. 2020). Further, the WHO declared the disease as Public Health Emergency of International Concern in January 2020 and officially named the disease caused by the novel CoV2 as coronavirus

Table 2.1 Major pandemic outbreaks in the history of mankind (Source: WHO, CDC USA)

S. No.	Pandemic disease outbreak	Organism responsible (virus/ bacteria/protozoan)	First outbreak/ start (year)	Number of deaths
1	Plague of Galen (Antonine Plague)	Measles/Variola virus	165 A.D.	~5 million
2	Bubonic plague (Plague of Justinian)	*Yersinia pestis*	540–542 A.D.	~25–50 million
3	Bubonic plague/Black Death	*Yersinia pestis*	1346	~200 million
4	Great Plague of Marseille	*Yersinia pestis*	1720	~1 million
5	Cholera	*Vibrio cholerae*	1817–1824	<1 million
6	Russian flu	Influenza A/H2N2/ H3N8	1889	<1 million
7	Spanish flu	Influenza A/H1N1	1918	20–50 million
8	Asian flu	Influenza A/H2N2	1958	<2 million
9	Hong Kong flu	Influenza A/H3N2	1968	<1 million
10	AIDS	HIV	1976	~36 million
11	SARS-CoV	Coronavirus	2002–2003	>1000

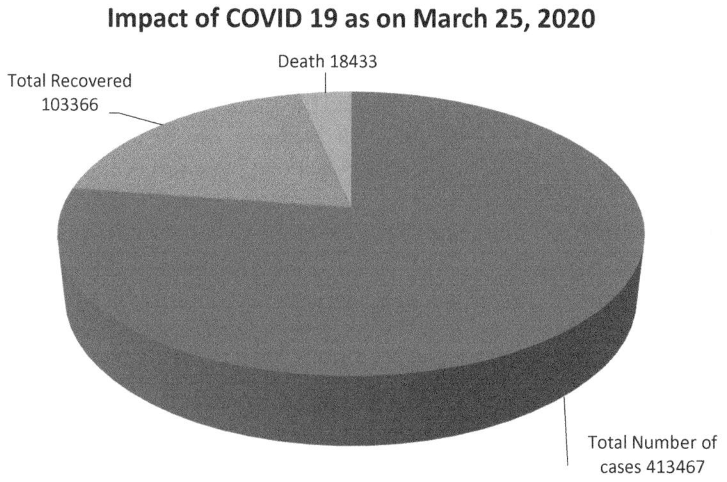

Fig. 2.1 (**a**) Worldwide reported cases, recovery, and death from COVID-19 as on March 25, 2020. (**b**) Worldwide increase in COVID-19 cases in 20 days (from March 25, 2020 to April 14, 2020)

disease 2019 (COVID-19) on February 12, 2020 (Adhikari et al. 2020; Zu et al. 2020). As per data available on various websites regarding COVID-19 infections worldwide, the cases are increasing exponentially. As on March 25, 2020, there were 413,467 reported cases, which included 18,433 deaths and 103,366 recovered cases which further rise 5× times to 18,48,439 including 117,217 deaths and 485,303 recovered cases world-wide till April 14, 2020. Figure 2.1a and b depicts the

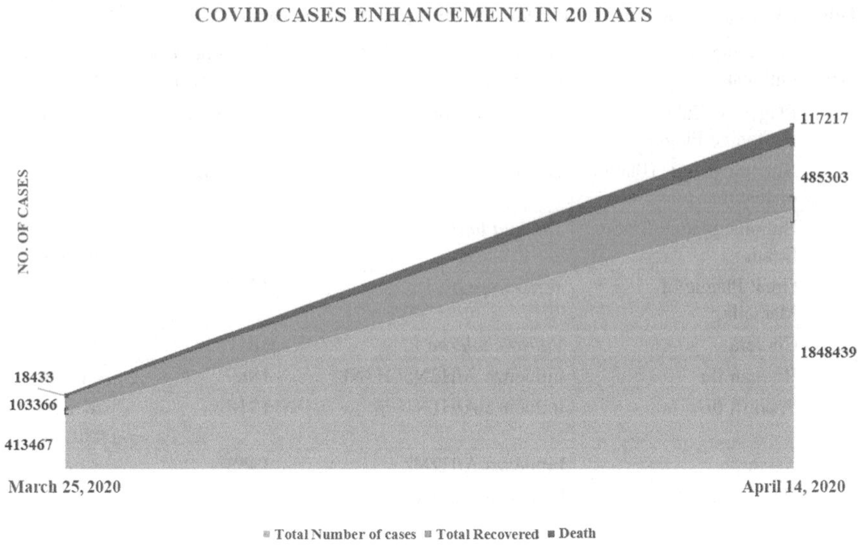

Fig. 2.1 (continued)

distribution of virus infection with death and recovery data as per the webpage worldometers.info.

The first 54 reported cases of COVID-19 were observed in December 2019 at Wuhan, China, and this has now spread across the globe affecting 440,318 people in 195 countries. The severity of the infection increases due to its human-to-human transmission ability majorly by means of contact and large droplets. Additionally, it can also be transmitted through aerosols and fomite on various surfaces and cause infection (Perlman 2020). This chapter intends to provide an insight on the epidemiology of COVID-19 for a better understanding of disease patterns and distribution.

2.2 Twenty-First Century Epidemics

The twenty-first century started with the SARS-CoV outbreak in China in the year 2002–2003. The disease spread rapidly throughout the world and infected approximately 8098 people in 37 countries and caused 778 deaths. The outbreak of the disease created fear worldwide (worldatlas.com). The Dengue virus on other hand appears to be big threat of severe level which infects approximately 400 million people worldwide each year and approximately 100 million of infected patients require critical care facility whereas around 22,000 lost their life due to dengue infection. After the SARS outbreak, there were a series of epidemics such as dengue, encephalitis, MERS, Zika, Ebola, Avian Flu, etc., that were reported from various parts of the world. Table 2.2 depicts the various epidemics of the twenty-first century as well as their occurrence and affected population (Source: WHO, CDC, worldpress.com) (Nag 2018).

Table 2.2 Epidemics of the twenty-first century

S.No.	Disease	Organism	Reported year	Number of people affected	Number of deaths
1	SARS-CoV	Coronavirus	2002–2003	8098	778
2	Zimbabwean cholera	*Vibrio cholerae*	2008	8500	4369
3	Flu	H1N1 Influenza A	2009		18,000
4	West African meningitis	*Neisseria meningitidis*	2009	13,516	931
5	Haitian cholera	*Vibrio cholerae*	2010	80,000	9985
6	Dengue fever outbreak	Dengue viruses	2011	21,204	<300
7	MERS	Coronavirus	2012	2494	858
8	Ebola	Ebola virus	2013	28,600	11,325
9	Zika	Zika virus	2013–2014	Approx. 2400	29 Babies in Brazil (2015)
10	Yemen cholera	*Vibrio cholerae*	2016	269,608	1614
11	Nipah	*Nipah virus*	2018	19	17
12	COVID-19	SARS-COV2	2019–2020	375,498 till March 25, 2020	16,362 till March 25, 2020

2.3 SARS-CoV-2 Outbreak

At the end of 2019, an unknown disease emerged and came into the spotlight. The disease caused pneumonia-like symptoms and lung fibrosis (Zhou et al. 2020). It emerged in the city of Wuhan, Hubei Province, China, which has a population of 11 million (Adhikari et al. 2020; Callaway et al. 2020; Chen et al. 2020; Fisher and Wilder-Smith 2020; Jingchun et al. 2020). China reported this pneumonia of unidentified cause first to the WHO country office on December 31, 2019 (WHO 2020). Since then, it has reported several thousand new cases of COVID-19. The peak of the epidemic in China was in late January and early February (Callaway et al. 2020). Up to January 31, 2020, COVID-19 had spread to 19 other countries, infecting 11,791 and causing 213 reported deaths. The COVID-19 epidemic was declared a Public Health Emergency of International Concern by the World Health Organization on January 30, 2020 (WHO 2020; Adhikari et al. 2020). It has since developed into a global pandemic and has affected huge numbers of people in Iran, South Korea, and Italy and has pushed a spike in worldwide cases across over 150 countries (Callaway et al. 2020).

On December 29, 2019, the WHO officially named the novel coronavirus as coronavirus disease 2019 (COVID-19). Currently, the virus is referred to as severe acute respiratory syndrome coronavirus 2 (SARS-CoV-2). According to reports, a number of people infected with pneumonia of unidentified cause were associated to a local seafood market in Wuhan, China, in December 2019. The Chinese Centre for Disease Control and Prevention (China CDC) immediately conducted epidemiological and etiological investigation. The WHO confirmed the association of the coronavirus outbreak with the seafood market of Wuhan (Sun et al. 2020). Immediately, scientists started to conduct research to find out the origin of the new coronavirus. The research group, led by Prof. Yong Zhang, were the first to publish the genome of COVID-19 on January 10, 2020 (Adhikari et al. 2020).

2.4 Pandemic SARS-CoV-2/COVID-19

Within a month of the outbreak at Wuhan, the SARS-CoV-2 virus extended rapidly all over China at the time of the Chinese New Year (Adhikari et al. 2020). The virus was not limited to a country. It was highly contagious and spread to more than 100 countries in the last 2–3 months and affected more than 300,000 people worldwide. As on March 24, 2020, the affected population is as follows: the Western Pacific Region under which China, Republic of Korea, Australia, Malaysia, Japan, Singapore, New Zealand, etc. come reported a total of 96,580 confirmed cases and 3502 deaths. On March 24, 2020, 943 new cases and 29 deaths were registered on a single day. The European Region (Italy, Spain, Germany, the United Kingdom, Norway, etc.) accounted for a total of 195,511 positive cases, out of which 24,087 were registered just in 1 day. The numbers peaked up to 10,189 confirmed cases and 1447 deaths in 1 day. In the Southeast Asia Region, 1990 confirmed cases were reported with 65 deaths. In the Eastern Mediterranean Region, a total of 27,215 people were affected and 1877 died due to this epidemic. In the Americas, 49,444 confirmed cases and 565 deaths were reported, with 12,428 new cases and 100 deaths registered in a day. Finally, in the African Region, 1305 confirmed cases and 26 deaths were reported. Table 2.3 provides a global scenario of the total number of COVID-19 positive cases and total number of deaths till March 25, 2020, taken from World health Organization situation report 2020.

As per WHO situation reports, the coronavirus started with a few positive cases but due to its highly contagious nature increased more than tenfold within 10 days' time.

In last 3 weeks, coronavirus disease has expanded rapidly across Europe, North America, Asia, and the Middle East, with the first confirmed cases identified in Latin American countries and African countries. Positive coronavirus cases outside China increased radically by March 16, 2020, and the number of affected countries, states, or territories reached 143 according to the WHO. Considering the alarming levels of infections and severity, the Director-General of WHO declared COVID-19 as a pandemic (Trevor Bedford et al. 2020). On March 13, 2020, the Director-General of the World Health Organization, Tedros Adhanom Ghebreyesus, said that Europe had become the epicenter of the pandemic (Trevor Bedford et al. 2020).

Figures 2.2 and 2.3 present the week-wise data of reported cases and deaths, respectively, in some of the majorly affected countries from January 21, 2020, to March 24, 2020,. Data has shown the exponential growth in the number of cases

Table 2.3 Globally confirmed COVID-19 positive cases and deaths from January 21, 2020, to March 25, 2020 (based on coronavirus disease (COVID-2019) situation reports)

WHO situation report	Globally confirmed cases	Globally reported deaths
21.01.2020	282	6
22.01.2020	314	6
23.01.2020	581	17
24.01.2020	846	25
25.01.2020	1320	41
26.01.2020	2014	56
27.01.2020	2798	80
28.01.2020	4593	106
29.01.2020	6065	132
30.01.2020	7818	170
31.01.2020	9826	213
1.02.2020	11,953	259
2.02.2020	14,557	305
3.02.2020	17,391	362
4.02.2020	20,630	426
5.02.2020	24,554	492
6.02.2020	28,276	565
7.02.2020	31,481	638
8.02.2020	34,886	724
9.02.2020	37,558	813
10.02.2020	40,554	910
11.02.2020	43,103	1018
12.02.2020	45,171	1115
13.02.2020	46,997	1369
14.02.2020	49,053	1383
15.02.2020	50,580	1526
16.02.2020	51,857	1669
17.02.2020	71,429	1775
18.02.2020	73,332	1873
19.02.2020	75,204	2009
20.02.2020	75,748	2129
21.02.2020	76,769	2247
22.02.2020	77,794	2359
23.02.2020	78,811	2462
24.02.2020	79,331	2618
25.02.2020	80,239	2700
26.02.2020	81,109	2762
27.02.2020	82,294	2804
28.02.2020	83,652	2858
29.02.2020	85,403	2924
1.03.2020	87,137	2977
2.03.2020	88,948	3043
3.03.2020	90,869	3112

(continued)

Table 2.3 (continued)

WHO situation report	Globally confirmed cases	Globally reported deaths
4.03.2020	93,091	3198
5.03.2020	95,324	3281
6.03.2020	98,192	3380
7.03.2020	101,927	3486
8.03.2020	105,586	3584
9.03.2020	109,577	3809
10.03.2020	113,702	4012
11.03.2020	118,319	4292
12.03.2020	125,260	4613
13.03.2020	132,758	4955
14.03.2020	142,534	5392
15.03.2020	153,517	5735
16.03.2020	167,515	6606
17.03.2020	179,111	7426
18.03.2021	191,127	7807
19.03.2020	209,839	8778
20.03.2020	234,073	9840
21.03.2020	266,073	11,183
22.03.2020	292,142	12,783
23.03.2020	332,930	14,509
24.03.2020	372,757	16,231
25.03.2020	413,467	18,433

World Health Organization, 2020. https://www.who.int/emergencies/diseases/novel-coronavirus-2019/situation-reports

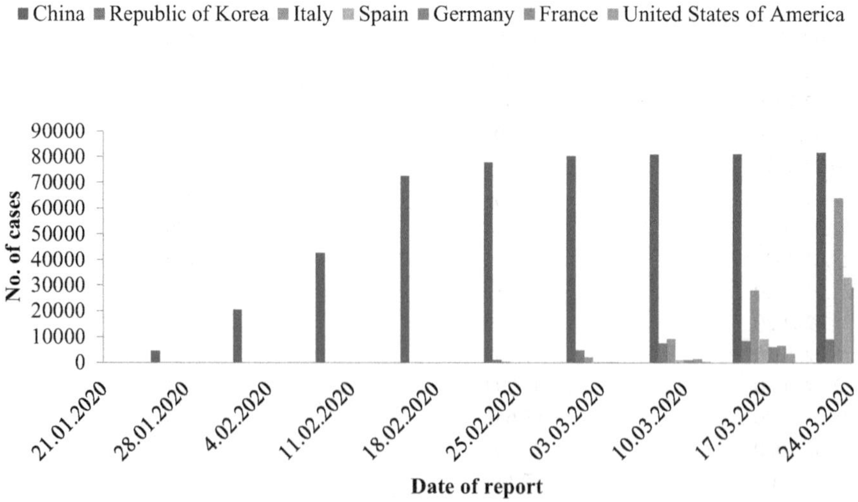

Fig. 2.2 Week-wise data of reported cases in some majorly affected countries (from January 21, 2020, to March 24, 2020). (Data taken from WHO situation reports)

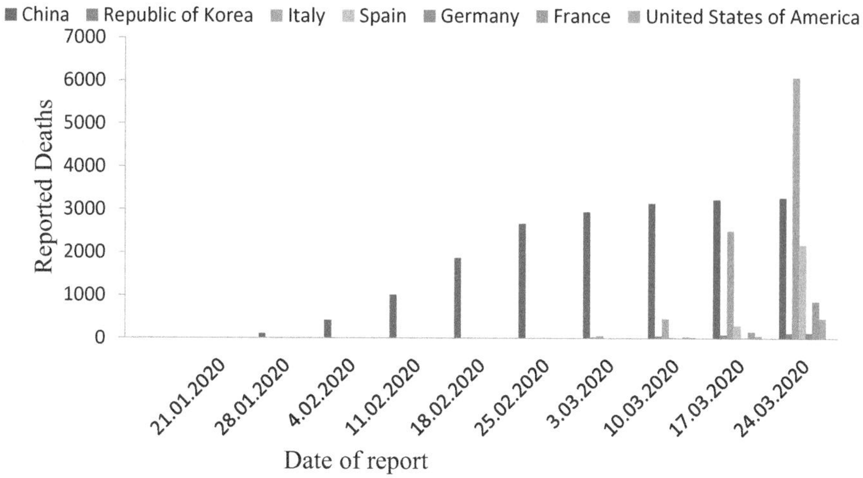

Fig. 2.3 Week-wise data of reported death cases in some majorly affected countries (from January 21, 2020, to March 24, 2020). (Data taken from WHO situation reports)

during the third week. China, Spain, Italy, and France all faced an explosion of cases in the third week.

2.5 Worldwide Surveillance Systems

As per the report published by WHO-China Joint Mission on Coronavirus Disease 2019 published on February 28, 2020, "The most of the global community is not ready for COVID 19." The disease emerged in China, and after several strict major decisions including lockdown to maintain social distancing, China successfully contained the virus. The number of new cases declined drastically (Fisher and Wilder-Smith 2020). The epicenter of the disease has now shifted to Europe, mainly to Italy and USA. The WHO is monitoring and tracking all the developments very closely around the world. The emergence of new cases as well as its spread to new countries and locations is being continuously tracked by the global monitoring system of the WHO. Apart from WHO surveillance, governments of various countries as well as their health ministries, home ministries, aviation, and non-government organizations are working together to keep a tab on the latest developments in COVID 19 cases. All the major airports of world were made fully equipped to scan and isolate passengers arriving from other countries or any other infected region.

Public health emergencies like COVID-19 require effective response in accurate time. Due to the insufficient information and suppress reporting on COVID19 outbreak from China many of the countries across the world were not able to understand the severity neither preprepared themselves to tackle this health emergency (Pisa 2020).

At national level, many governments developed a digital surveillance system to keep track of international travelers, isolated people, COVID-19 positive contacted people, and those with quarantine status. This close monitoring will help achieve a decline in COVID-19 cases as well as in containment of the virus. Additionally, several governments have developed online live digital tools to continuously monitor and update the current numbers of COVID-19 cases. The WHO live situation reports provide live status of emerging cases across the globe and pattern of disease spread. Besides this, some other independent agencies are also keeping close watch on COVID-19 development such as John Hopkins University, Centre for Disease Control, worldometer.info, NIH, etc. The Pandemicity project, a multilingual medical research database for nCoV-19 started by pandemicity.org, includes several tools for screening and calculating risk factors of COVID-19 online. Many scientific publication houses are providing open access to COVID-19-related research works, which enables researchers worldwide to develop effective diagnostic and therapeutic solutions. The effective surveillance from local community level to regional level to national level to world level is the key to find effective solutions for containment of the virus. Worldwide surveillance also provides on-time alerts as to potential locations where the infection may be emerging. Additionally surveillance enables enhanced cooperation among affected countries for formulation of collective measures to fight pandemic like COVID19.

2.6 Regions of Potential Threat

The disease emerged from Wuhan, Hubei Province, China, as an unknown viral pneumonia diagnosed in 54 individual and by March 25, 2020, it had infected 81,848 people in China alone. Wuhan became the epicenter of COVID-19 just after emergence of the disease. Government and local bodies worked tirelessly to control the spread of COVID-19. Initially COVID-19 was not identified as a human–to-human transmission disease, but with exponential increase in cases research on COVID-19 established the fact that disease is transmitted from human to human via aerosol. The SARS-CoV-2 virus is a novel coronavirus of the family Coronaviridae and very little is known about its characteristics (Tekes and Thiel 2016; Ashour et al. 2020). As we gain more knowledge about CoV2, its spread is also increasing exponentially. The initial epicenter of the disease shifted from Wuhan to Europe to the USA. Italy appears worst affected having 69,176 reported cases till March 25, 2020, with the maximum number of deaths of 6820 infected people. USA with 51,914 positive cases is emerging as another epicenter. Since there is no vaccine or medication available for the novel coronavirus, social distancing and proper maintenance of self as well as surrounding sanitation is the only treatment. The long and strict lockdown in China shows excellent decline in viral infection and social distancing found very helpful in containment of the virus. The initial lighter attitude of European countries and the USA is one of the major reasons for their emergence as an epicenter of the disease. Highly populated countries like India must learn from the mistakes of these nations and implement strict measures for containment of the virus, as uncontrolled outbreak of the disease could lead to a devastating situation for developing countries like India. Another region of potential threat is Iran where the number of deaths has increased exponentially, with 2077 deaths recorded till March 25, 2020. Underdeveloped countries with lack of basic facilities, public awareness, and poverty like Pakistan and African countries may emerge as another epicenter for COVID-19. Special attention and extended help need to be given to such nations for fighting the pandemic.

Executive Summary

- 416,686 people were infected with SARS-CoV-2 till March 25, 2020 including 18,589 deaths.
  - COVID-19 emerged from China and has now spread to 196 countries worldwide.
- The disease has been declared a Public Health Emergency of International Concern by WHO and as a global pandemic.
  - On February 12, 2020, WHO renamed the novel CoV COVID-19.
- SARS-CoV-2 shows genetic similarity to SARS CoV and Bat CoV.
- Europe and USA have emerged as the new epicenters for COVID-19.
  - World level surveillance with minute-by-minute tracking of COVID-19 is being performed by the WHO as well as some other government and private agencies.
- COVID-19 is a novel virus and proper precaution is the only cure.
  - Maintenance of social distancing and proper hygiene is the only cure of COVID-19 till any vaccine is developed and made available.

2.7 Conclusions

SARS-CoV-2 is a highly infectious virus with the ability of human-to-human transmission. The virus is of zoonotic origin and was first transmitted into humans from animal (Menachery et al. 2018). SARS-CoV-2 has now spread to 196 countries, raising infection exponentially worldwide. Active worldwide surveillance is continuously being done by the WHO to monitor new cases, deaths, and recoveries. The mortality rate is quite low, ranging between 1.2 and 14%, depending upon several parameters such as age, health condition, immunity, diabetes, and past disease records. International travel needs to be avoided to potential regions like Italy, USA, and China. Safety measures such as social distancing and maintenance of hygiene is top priority to secure self from COVID-19 infection.

2.8 Future Perspectives

The epidemiology of COVID-19 provides a better understanding about the pattern of disease growth and spread. The high potential of COVID-19 causing infection from human to human may be neutralized by following social distancing and sanitization. Governments may impose lockdown and curfew for strict implementation of social distancing to reduce the spread of the infection. Worldwide surveillance system needs to be improved by use of artificial intelligence and information technology.

References

Adhikari SP, Meng S, Wu Y-J, Mao Y-P, Ye R-X, Wang Q-Z, Sun C, Sylvia S, Rozelle S, Raat H, Zhou H (2020) Epidemiology, causes, clinical manifestation and diagnosis, prevention and control of coronavirus disease (COVID-19) during the early outbreak period: a scoping review. Infect Dis Poverty 9(1):29

Ashour MH, Elkhatib FW, Rahman MM, Elshabrawy AH (2020) Insights into the recent 2019 novel coronavirus (SARS-CoV-2) in light of past human coronavirus outbreaks. Pathogens 9(3): E186

Bai Y, Yao L, Wei T, Tian F, Jin D-Y, Chen L, Wang M (2020) Presumed asymptomatic carrier transmission of COVID-19. JAMA 323(14):1406–1407

Callaway E, Cyranoski D, Mallapaty S, Stove E, Tollefson J (2020) The coronavirus pandemic in five powerful charts. Nature 579:482–483

Chen N, Zhou M, Dong X, Qu J, Gong F, Han Y, Qiu Y, Wang J, Liu Y, Wei Y, Xia JA, Yu T, Zhang X, Zhang L (2020) Epidemiological and clinical characteristics of 99 cases of 2019 novel coronavirus pneumonia in Wuhan, China: a descriptive study. Lancet 395(10223):507–513

Fisher D, Wilder-Smith A (2020) The global community needs to swiftly ramp up the response to contain COVID-19. Lancet 395(10230):1109–1110

Jingchun F, Xiaodong L, Weimin P, Mark WD, Shisan B (2020) Epidemiology of 2019 novel coronavirus disease-19 in Gansu Province, China, 2020. Emerg Infect Dis J 26(6). https://doi.org/10.3201/eid2606.200251

Lamb R (2013) 10 worst epidemics. https://www.seeker.com/10-worst-epidemics-1767852043.html

Lu R, Zhao X, Li J, Niu P, Yang B, Wu H, Wang W, Song H, Huang B, Zhu N, Bi Y, Ma X, Zhan F, Wang L, Hu T, Zhou H, Hu Z, Zhou W, Zhao L, Chen J, Meng Y, Wang J, Lin Y, Yuan J, Xie Z, Ma J, Liu WJ, Wang D, Xu W, Holmes EC, Gao GF, Wu G, Chen W, Shi W, Tan W (2020) Genomic characterisation and epidemiology of 2019 novel coronavirus: implications for virus origins and receptor binding. Lancet 395(10224):565–574

Menachery VD, Gralinski LE, Mitchell HD, Dinnon KH, Leist SR, Yount BL, McAnarney ET, Graham RL, Waters KM, Baric RS (2018) Combination attenuation offers strategy for live attenuated coronavirus vaccines. J Virol 92(17):e00710–e00718

Nag OS (2018) The deadliest epidemics of the 21st century so far. worldatlas.com/articles/the-deadliest-epidemics-of-the-21st-century-till-date.html

Perlman S (2020) Another decade, another coronavirus. N Engl J Med 382(8):760–762

Pisa M (2020) COVID-19, information problems, and digital surveillance. https://www.cgdev.org/blog/covid-19-information-problems-and-digital-surveillance

Staff (2020) Outbreak: 10 of the worst pandemics in history. https://www.mphonline.org/worst-pandemics-in-history/

Sun K, Chen J, Viboud C (2020) Early epidemiological analysis of the coronavirus disease 2019 outbreak based on crowd sourced data: a population-level observational study. Lancet Digit Health 2(4):e201–e208

Tekes G, Thiel HJ (2016) Chapter six—feline coronaviruses: pathogenesis of feline infectious peritonitis. In: Ziebuhr J (ed) Advances in virus research, vol 96. Academic Press, New York, pp 193–218

Trevor Bedford RN, Hadfield J, Hodcroft E, Ilcisin M, Müller N (2020) Genomic analysis of nCoV spread. Situation report 2020-01-25. https://nextstrain.org/narratives/ncov/sit-rep/2020-01-25

WHO (2020) COVID19: rolling updates on coronavirus disease (COVID-19). https://www.who.int/emergencies/diseases/novel-Coronavirus-2019/events-as-they-happen

Zhou F, Yu T, Du R, Fan G, Liu Y, Liu Z, Xiang J, Wang Y, Song B, Gu X, Guan L, Wei Y, Li H, Wu X, Xu J, Tu S, Zhang Y, Chen H, Cao B (2020) Clinical course and risk factors for mortality of adult inpatients with COVID-19 in Wuhan, China: a retrospective cohort study. Lancet 395 (10229):1054–1062

Zu ZY, Jiang MD, Xu PP, Chen W, Ni QQ, Lu GM, Zhang LJ (2020) Coronavirus disease 2019 (COVID-19): a perspective from China. Radiology 200490

Chapter 3
Morphology, Genome Organization, Replication, and Pathogenesis of Severe Acute Respiratory Syndrome Coronavirus 2 (SARS-CoV-2)

Swatantra Kumar, Rajni Nyodu, Vimal K. Maurya, and Shailendra K. Saxena

Abstract SARS-CoV-2 is a single-stranded RNA virus of ~30 kb genome size which belongs to genus *Coronavirus* and family *Coronaviridae*. SARS-CoV-2 has recently emerged and has been declared as a pandemic by the World Health Organization. Genomic characterization of SARS-CoV-2 has shown that it is of zoonotic origin. The structure of SARS-CoV-2 is found to be similar to SARS-CoV with virion size ranging from 70 to 90 nm. Spike, membrane, and envelope surface viral proteins of coronavirus are embedded in host membrane-derived lipid bilayer encapsulating the helical nucleocapsid comprising viral RNA. The genome comprises of 6–11 open reading frames (ORFs) with 5′ and 3′ flanking untranslated regions (UTRs). Sequence variation among SARS-CoV-2 and SARS-CoV revealed no significant difference in ORFs and nsps. The nsps includes two viral cysteine proteases including papain-like protease (nsp3), chymotrypsin-like, 3C-like, or main protease (nsp5), RNA-dependent RNA polymerase (nsp12), helicase (nsp13), and others likely to be involved in the transcription and replication of SARS-CoV-2. The structure of spike glycoprotein structure of SARS-CoV-2 resembles that of the spike protein of SARS-CoV with an root-mean-square deviation (RMSD) of 3.8 Å. Like SARS-CoV, SARS-CoV-2 uses the ACE2 receptor for internalization and TMPRSS2 serine proteases for S protein priming. Histopathological investigation of tissues from SARS-CoV-2 infected patients showed virus-induced cytopathic effect with signs of acute respiratory distress syndrome in lung cells. This chapter discusses about the morphology, genome organization, replication, and pathogenesis

S. Kumar · R. Nyodu · V. K. Maurya
Centre for Advanced Research (CFAR), Faculty of Medicine, King George's Medical University (KGMU), Lucknow, India

S. K. Saxena (✉)
Centre for Advanced Research (CFAR)-Stem Cell/Cell Culture Unit, Faculty of Medicine, King George's Medical University (KGMU), Lucknow, India
e-mail: shailen@kgmcindia.edu

S. K. Saxena (ed.), *Coronavirus Disease 2019 (COVID-19)*, Medical Virology: from Pathogenesis to Disease Control, https://doi.org/10.1007/978-981-15-4814-7_3

of SARS-CoV-2 that may help us understand the disease that may leads to identification of effective antiviral drugs and vaccines.

Keywords SARS-CoV-2 · Genome · Spike glycoprotein · ACE2 · Entry · Replication · Pathogenesis

3.1 Introduction

Coronaviruses (CoVs) are enveloped single-stranded positive sense RNA viruses that belong to the family *Coronaviridae*. On the basis of genomic organization and phylogenetic relationship, coronaviruses have been classified into the subfamily *Coronavirinae* that consists of four genera *Alphacoronavirus* (αCoV), *Betacoronavirus* (βCoV), *Gammacoronavirus* (γCoV), and *Deltacoronavirus* (δCoV) (Cui et al. 2019). Evolutionary trend analysis of coronaviruses has revealed that αCoV and βCoV originated from bats and rodents, while γCoV and δCoV were found to have originated from avian species (Ge et al. 2017). The ability of CoVs to cross the species barrier has resulted in some of the pathogenic CoVs. HKU1, NL63, OC43, and 229E CoVs are associated with mild symptoms in humans, whereas severe acute respiratory syndrome CoV (SARS-CoV) and Middle East respiratory syndrome CoV (MERS-CoV) are known to cause severe disease (Fehr and Perlman 2015). In 2002–2003, SARS-CoV emerged in China with 8000 clinical cases and 800 deaths. Since 2012, MERS-CoV has caused persistent epidemics in the Arabian Peninsula. Both the viruses have been found to originate from bats and then transmitted into intermediate mammalian host civets in the case of SARS-CoV and camels in the case of MERS-CoV and eventually infected humans (Song et al. 2019).

SARS-CoV-2 has been declared as a pandemic, with 1,844,683 confirmed cases and 117,021 deaths globally by 14th April 2020 (World Health Organization 2020). To characterize the novel coronavirus, bronchoalveolar lavage fluid and throat swabs were collected from nine patients who had visited the Wuhan seafood market during the initial outbreak. Special pathogen-free human airway epithelial (HAE) cells were used for virus isolation. The collected samples were inoculated into the HAE cells through the apical surfaces. HAE cells were monitored for cytopathic effects and supernatant was collected to perform RT-PCR assays. Apical samples were collected for next generation sequencing after three passages. The whole-genome sequences of SARS-CoV-2 were generated by a combination of Sanger, Illumina, and Oxford nanopore sequencing (Lu et al. 2020). Phylogenetic analysis has revealed that bats might be at the source of SARS-CoV-2 (Andersen et al. 2020). Additionally, some studies have suggested that the origin of SARS-CoV-2 is associated with pangolins (Li et al. 2020; Shereen et al. 2020). To decipher the mechanism of replication and development of effective preventive and therapeutic strategies, understanding the structure of SARS-CoV-2, genome organization, and replication is crucial. Therefore this chapter focuses on the morphology and structure, genomic organization, and replication cycle of SARS-CoV-2.

3.2 Morphology of SARS-CoV-2

SARS-CoV-2 isolated from nasopharyngeal and oropharyngeal samples were inoculated on the vero cells. In order to identify SARS-CoV-2, inoculated cells were prefixed using 2% paraformaldehyde and 2.5% glutaraldehyde, and transmission electron microscopy was performed. The structure of SARS-CoV-2 observed by examining infected cells after 3 days post infection. Electron microscopy revealed the coronavirus-specific morphology of SARS-CoV-2 with virus particle sizes ranging from 70 to 90 nm observed under a wide variety of intracellular organelles, most specifically in vesicles (Park et al. 2020). Due to high sequence similarity, the structure of SARS-CoV-2 is speculated to be the same as SARS-CoV (Kumar et al. 2020). The surface viral protein spike, membrane, and envelope of coronavirus are embedded in host membrane-derived lipid bilayer encapsulating the helical nucleocapsid comprising viral RNA (Fig. 3.1) (Finlay et al. 2004). The structure of spike (Yan et al. 2020) and protease of SARS-CoV-2 (Zhang et al. 2020) has been resolved, which provides an opportunity to develop a newer class of drugs for treatment of COVID-19.

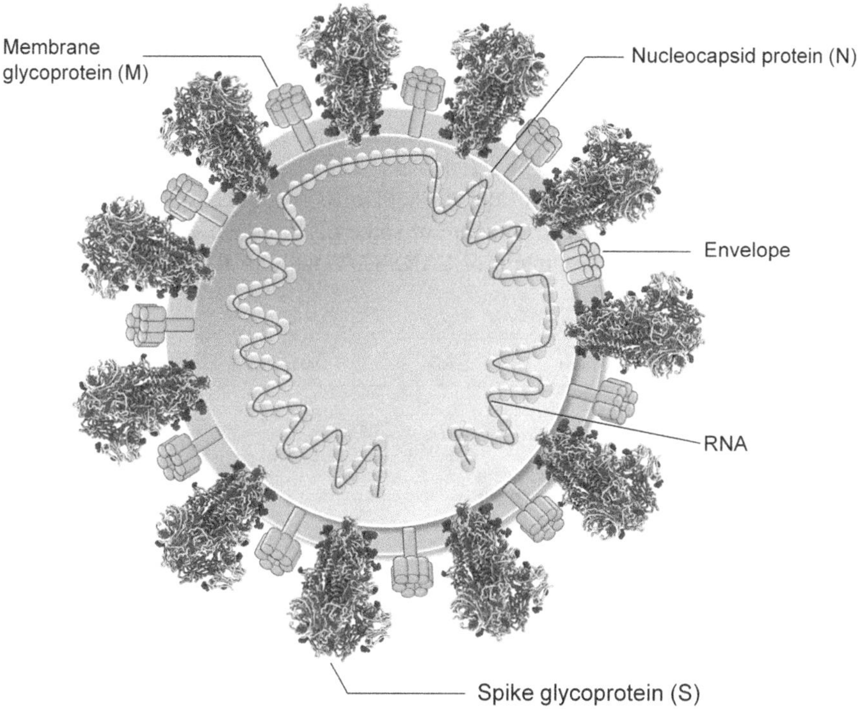

Fig. 3.1 Structure of SARS-CoV-2. SARS-CoV-2 has surface viral proteins, namely, spike glycoprotein (S), which mediates interaction with cell surface receptor ACE2. The viral membrane glycoprotein (M) and envelope (E) of SARS-CoV-2 are embedded in host membrane-derived lipid bilayer encapsulating the helical nucleocapsid comprising viral RNA

3.3 Genome Organization of SARS-CoV-2

The size of coronavirus genome is in the range of 26 to 32 kb and comprise 6–11 open reading frames (ORFs) encoding 9680 amino acid polyproteins (Guo et al. 2020). The first ORF comprises approximately 67% of the genome that encodes 16 nonstructural proteins (nsps), whereas the remaining ORFs encode for accessory and structural proteins. The genome of SARS-CoV-2 lacks the hemagglutinin-esterase gene. However, it comprises two flanking untranslated regions (UTRs) at 5′ end of 265 and 3′ end of 358 nucleotides. Sequence variation among SARS-CoV-2 and SARS-CoV revealed no significant difference in ORFs and nsps. The nsps includes two viral cysteine proteases including papain-like protease (nsp3), chymotrypsin-like, 3C-like, or main protease (nsp5), RNA-dependent RNA polymerase (nsp12), helicase (nsp13), and others likely to be involved in the transcription and replication of SARS-CoV-2 (Chan et al. 2020). In addition to nsps, four major structural proteins are spike surface glycoprotein (S), membrane, nucleocapsid protein (N), envelope (E) and accessory proteins encoded by ORFs. N-terminal glycosylated ectodomain is present at the N-terminal end of M protein that comprises of three transmembrane domains (TM) and a long C-terminal CT domain (Fig. 3.2).

The M and E proteins are required for virus morphogenesis, assembly, and budding, whereas S glycoprotein is a fusion viral protein comprising two subunits S1 and S2. The S1 subunit, which shares 70% sequence identity with bat SARS-like CoVs and human SARS-CoV, comprises signal peptide, N-terminal domain (NTD), and receptor-binding domain (RBD) (Walls et al. 2020). Most of the differences were found in the external subdomain that is primarily responsible for interaction of spike with the ACE2 receptor. The ectodomain of spike protein (1–1208 amino acid residues) was cloned, expressed and crystallize to solve the spike glycoprotein structure of SARS-CoV-2. The structure of spike glycoprotein structure of SARS-CoV-2 resembles the spike protein of SARS-CoV with an RMSD of 3.8 Å. The

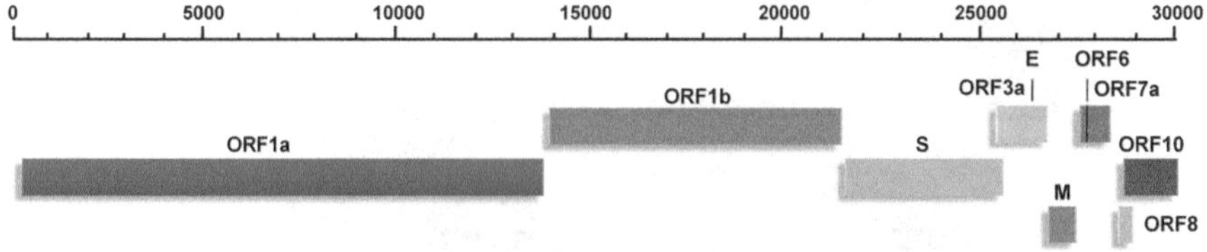

Fig. 3.2 Genomic organization of SARS-CoV-2. The size of coronavirus genome ranges from 26 to 32 kb and comprises 6–11 open reading frames (ORFs) encoding 9680 amino acid polyprotein. The first ORF comprises of approximately 67% of the genome that encodes 16 nonstructural proteins (nsps), whereas the remaining ORFs encode for accessory and structural proteins. The nsps includes two viral cysteine proteases, including papain-like protease (nsp3), chymotrypsin-like, 3C-like, or main protease (nsp5), RNA-dependent RNA polymerase (nsp12), helicase (nsp13), and others likely to be involved in the transcription and replication of SARS-CoV-2. In addition to nsps, the genome encodes for four major structural proteins including spike surface glycoprotein (S), membrane, nucleocapsid protein (N), envelope (E) and accessory proteins like ORFs

study also reveals that the receptor-binding region (RBD) exhibited the highest structural divergence (Wrapp et al. 2020). The S2 subunit that shares 99% sequence identity with bat SARS-like CoVs and human SARS-CoV comprises two heptad repeat regions known as HR-N and HR-C, which form the coiled-coil structures surrounded by the protein ectodomain. The S protein has been found to exhibit a furin cleavage site (PRRARS'V) at the interface between S1 and S2 subunits that is processed during the biogenesis (Coutard et al. 2020).

3.4 Entry and Replication of SARS-CoV-2 in Host Cells

Entry of coronaviruses into host target cells depends on the binding of spike glycoprotein to the cellular receptor and priming of S protein by host cell proteases. Like SARS-CoV, SARS-CoV-2 uses the ACE2 receptor for internalization and TMPRSS2 serine proteases for S protein priming (Hoffmann et al. 2020). Similar to SARS-CoV, the extrapulmonary spread of SARS-CoV-2 may be seen due to the widespread tissue expression of the ACE2 receptor. In addition, studies revealed that the spike protein of SARS-CoV-2 exhibits 10–20 times higher affinity as compared to that of SARS-CoV (Wrapp et al. 2020). Binding of spike protein to the ACE2 receptor results in conformational changes in spike protein that leads to the fusion of viral envelop protein with host cell membrane following entry via endosomal pathway (Coutard et al. 2020; Matsuyama and Taguchi 2009). This event is followed by the release of viral RNA into the host cytoplasm that undergoes translation and generates replicase polyproteins pp1a and pp1b that further cleaved by virus encoded proteinases into small proteins. The replication of coronavirus involves ribosomal frame shifting during the translation process and generates both genomic and multiple copies of subgenomic RNA species by discontinuous transcription that encodes for relevant viral proteins. Assembly of virion takes place via interaction of viral RNA and protein at endoplasmic reticulum (ER) and Golgi complex. These virions are subsequently released out of the cells via vesicles (Fig. 3.3) (Hoffmann et al. 2020).

3.5 Pathogenesis of SARS-CoV-2

The pathological findings of SARS-CoV-2 infected patients highly resemble that of SARS-CoV and MERS-CoV infected patients. Flow cytometric analysis of peripheral blood samples showed significant reduction of CD4 and CD8 T cell counts, and their status was found to be hyperactivated as higher proportion of dual positive (HLA-DR and CD38) was seen. Rapid progression of pneumonia was seen in chest X-ray images with some differences between the right and left lung. Histopathological investigation of lung, liver, and heart tissue was performed. Lung biopsy showed cellular fibromyxoid exudates with bilateral diffuse alveolar damage. The right lung

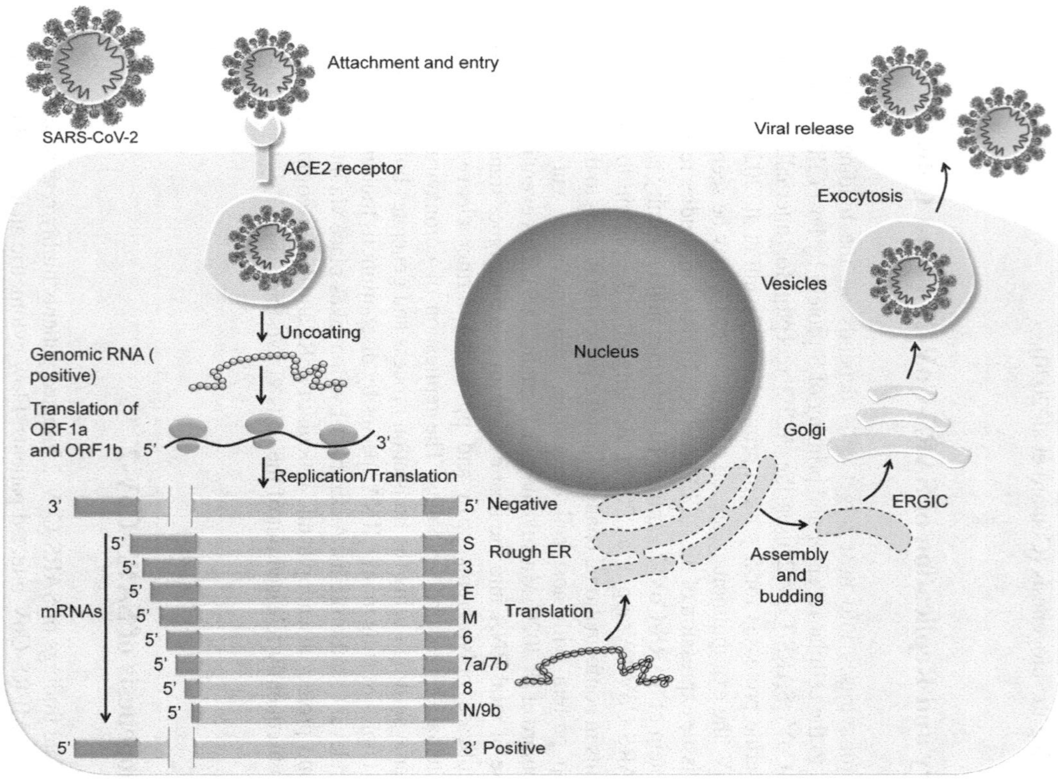

Fig. 3.3 Entry and replication of SARS-CoV-2 in host cells. Entry of SARS-CoV-2 into host target cells depends on the binding of spike glycoprotein to the cellular receptor ACE2 for internalization. Internalization results in uncoating of viral RNA into cytoplasm that undergoes translation and generates replicase polyproteins pp1a and pp1b, which is further cleaved by virus-encoded proteinases into small proteins. The replication of SARS-CoV-2 involves ribosomal frame shifting during the translation process and generates both genomic and multiple copies of subgenomic RNA species by discontinuous transcription required for relevant viral proteins. Assembly of virion takes place via interaction of viral RNA and protein at endoplasmic reticulum (ER) and Golgi complex. These virions are subsequently released out of the cells via vesicles via exocytosis

showed prominent desquamation of pneumocytes and formation of hyaline membrane, indicating signs of acute respiratory distress syndrome (ARDS), whereas the left lung showed pulmonary edema with formation of hyaline membrane (Xu et al. 2020). In addition, both lungs were found to exhibit interstitial mononuclear patchy inflammatory infiltrates dominated specifically by lymphocytes (Tian et al. 2020). The intra-alveolar spaces were characterized by multinucleated syncytial cells with atypical enlarged pneumocytes showing virus-induced cytopathic effect. Liver biopsy of patients infected with SARS-CoV-2 showed moderate microvesicular steatosis and mild portal and lobular activity, suggesting that injury might have been caused by the virus or drug induced. A few interstitial mononuclear inflammatory infiltrates were observed in the heart tissue. These pathological changes may provide new insights into the pathogenesis of pneumonia induced by SARS-CoV-2 that may help clinicians to effectively deal with COVID-19 patients.

3.6 Conclusions

Phylogenetic analysis revealed that SARS-CoV-2 might have originated from bats or pangolins. Structural investigations of virus-infected cells reveal the coronavirus-specific morphology of SARS-CoV-2 and the size of the virus (70–90 nm). The size of SARS-CoV-2 genome ranges from 26 to 32 kb and comprises 6–11 ORFs which lacks hemagglutinin-esterase gene. However, it comprises of 5′ and 3′ flanking untranslated regions (UTRs). The spike glycoprotein structure of SARS-CoV-2 resembles the spike protein of SARS-CoV with an RMSD of 3.8 Å. Like SARS-CoV, SARS-CoV-2 uses the ACE2 receptor for internalization and TMPRSS2 serine proteases for S protein priming. Histopathological investigation of tissues from SARS-CoV-2 infected patients showed virus-induced cytopathic effect with signs of acute respiratory distress syndrome in lung cells.

3.7 Future Perspectives

SARS-CoV-2 has recently emerged and has been declared as a pandemic by the World Health Organization. Based on the genomic sequences submitted to NCBI database, the scientific community has analyzed the samples and suggested preventive and therapeutic strategies. Therefore, investigation of genomic diversity in the collected specimens from around the globe needs to be conducted in order to design common, effective therapies and vaccines. In addition, genomic characterization helps us accurately identify the origin and evolution of the virus. Deciphering the mechanism of SARS-CoV-2 replication in various cell-based models may help us understand the pathogenesis and identify specific targets to develop effective antiviral drugs.

Acknowledgments The authors are grateful to the Vice Chancellor, King George's Medical University (KGMU), Lucknow, India, for the encouragement of this work. The authors have no other relevant affiliations or financial involvement with any organization or entity with a financial interest in or financial conflict with the subject matter or materials discussed in the manuscript apart from those disclosed.

References

Andersen KG, Rambaut A, Lipkin WI et al (2020) The proximal origin of SARS-CoV-2. Nat Med 26(4):450–452. https://doi.org/10.1038/s41591-020-0820-9

Chan JF, Kok KH, Zhu Z, Chu H, To KK, Yuan S, Yuen KY (2020) Genomic characterization of the 2019 novel human-pathogenic coronavirus isolated from a patient with atypical pneumonia after visiting Wuhan. Emerg Microbes Infect 9(1):221–236. https://doi.org/10.1080/22221751.2020.1719902. eCollection 2020

Coutard B, Valle C, de Lamballerie X, Canard B, Seidah NG, Decroly E (2020) The spike glycoprotein of the new coronavirus 2019-nCoV contains a furin-like cleavage site absent in CoV of the same clade. Antivir Res 176:104742. https://doi.org/10.1016/j.antiviral.2020.104742

Cui J, Li F, Shi ZL (2019) Origin and evolution of pathogenic coronaviruses. Nat Rev Microbiol 17 (3):181–192. https://doi.org/10.1038/s41579-018-0118-9

Fehr AR, Perlman S (2015) Coronaviruses: an overview of their replication and pathogenesis. Methods Mol Biol 1282:1–23. https://doi.org/10.1007/978-1-4939-2438-7_1

Finlay BB, See RH, Brunham RC (2004) Rapid response research to emerging infectious diseases: lessons from SARS. Nat Rev Microbiol 2(7):602–607

Ge XY, Yang WH, Zhou JH, Li B, Zhang W, Shi ZL, Zhang YZ (2017) Detection of alpha- and betacoronaviruses in rodents from Yunnan, China. Virol J 14(1):98. https://doi.org/10.1186/s12985-017-0766-9

Guo YR, Cao QD, Hong ZS, Tan YY, Chen SD, Jin HJ, Tan KS, Wang DY, Yan Y (2020) The origin, transmission and clinical therapies on coronavirus disease 2019 (COVID-19) outbreak—an update on the status. Mil Med Res 7(1):11. https://doi.org/10.1186/s40779-020-00240-0

Hoffmann M, Kleine-Weber H, Schroeder S, Krüger N, Herrler T, Erichsen S, Schiergens TS, Herrler G, Wu NH, Nitsche A, Müller MA, Drosten C, Pöhlmann S (2020) SARS-CoV-2 cell entry depends on ACE2 and TMPRSS2 and is blocked by a clinically proven protease inhibitor. Cell 181:1–10. https://doi.org/10.1016/j.cell.2020.02.052. [Epub ahead of print]

Kumar S, Maurya VK, Prasad AK et al (2020) Structural, glycosylation and antigenic variation between 2019 novel coronavirus (2019-nCoV) and SARS coronavirus (SARS-CoV). VirusDis 31(1):13–21. https://doi.org/10.1007/s13337-020-00571-5

Li X, Zai J, Zhao Q, Nie Q, Li Y, Foley BT, Chaillon A (2020) Evolutionary history, potential intermediate animal host, and cross-species analyses of SARS-CoV-2. J Med Virol 2020:1–10. https://doi.org/10.1002/jmv.25731

Lu R, Zhao X, Li J, Niu P, Yang B, Wu H, Wang W, Song H, Huang B, Zhu N, Bi Y, Ma X, Zhan F, Wang L, Hu T, Zhou H, Hu Z, Zhou W, Zhao L, Chen J, Meng Y, Wang J, Lin Y, Yuan J, Xie Z, Ma J, Liu WJ, Wang D, Xu W, Holmes EC, Gao GF, Wu G, Chen W, Shi W, Tan W (2020) Genomic characterisation and epidemiology of 2019 novel coronavirus: implications for virus origins and receptor binding. Lancet 395(10224):565–574. https://doi.org/10.1016/S0140-6736(20)30251-8

Matsuyama S, Taguchi F (2009) Two-step conformational changes in a coronavirus envelope glycoprotein mediated by receptor binding and proteolysis. J Virol 83(21):11133–11141. https://doi.org/10.1128/JVI.00959-09

Park WB, Kwon NJ, Choi SJ, Kang CK, Choe PG, Kim JY, Yun J, Lee GW, Seong MW, Kim NJ, Seo JS, Oh MD (2020) Virus isolation from the first patient with SARS-CoV-2 in Korea. J Korean Med Sci 35(7):e84. https://doi.org/10.3346/jkms.2020.35.e84

Shereen MA, Khan S, Kazmi A, Bashir N, Siddique R (2020) COVID-19 infection: origin, transmission, and characteristics of human coronaviruses. J Adv Res 24:91–98. https://doi.org/10.1016/j.jare.2020.03.005

Song Z, Xu Y, Bao L, Zhang L, Yu P, Qu Y, Zhu H, Zhao W, Han Y, Qin C (2019) From SARS to MERS, thrusting coronaviruses into the spotlight. Viruses. 11(1). pii: E59. https://doi.org/10.3390/v11010059

Tian S, Hu W, Niu L, Liu H, Xu H, Xiao SY (2020) Pulmonary pathology of early-phase 2019 novel coronavirus (COVID-19) pneumonia in two patients with lung cancer. J Thorac Oncol. pii: S1556-0864(20)30132-5. https://doi.org/10.1016/j.jtho.2020.02.010. [Epub ahead of print]

Walls AC, Park YJ, Tortorici MA, Wall A, McGuire AT, Veesler D (2020) Structure, function, and antigenicity of the SARS-CoV-2 spike glycoprotein. Cell 180:1–12. https://doi.org/10.1016/j.cell.2020.02.058

World Health Organization (2020) Coronavirus disease 2019 (COVID-19) situation report—85. World Health Organization. https://www.who.int/docs/default-source/coronaviruse/situation-reports/20200414-sitrep-85-covid-19.pdf?sfvrsn=7b8629bb_4. Accessed 14 Apr 2020

Wrapp D, Wang N, Corbett KS, Goldsmith JA, Hsieh CL, Abiona O, Graham BS, McLellan JS (2020) Cryo-EM structure of the 2019-nCoV spike in the prefusion conformation. Science 367 (6483):1260–1263. https://doi.org/10.1126/science.abb2507

Xu Z, Shi L, Wang Y, Zhang J, Huang L, Zhang C, Liu S, Zhao P, Liu H, Zhu L, Tai Y, Bai C, Gao T, Song J, Xia P, Dong J, Zhao J, Wang FS (2020) Pathological findings of COVID-19 associated with acute respiratory distress syndrome. Lancet Respir Med 8(4):420–422. https://doi.org/10.1016/S2213-2600(20)30076-X. [Epub ahead of print]

Yan R, Zhang Y, Li Y, Xia L, Guo Y, Zhou Q (2020) Structural basis for the recognition of the SARS-CoV-2 by full-length human ACE2. Science 367(6485):1444–1448. https://doi.org/10.1126/science.abb2762

Zhang L, Lin D, Sun X, Curth U, Drosten C, Sauerhering L, Becker S, Rox K, Hilgenfeld R (2020) Crystal structure of SARS-CoV-2 main protease provides a basis for design of improved α-ketoamide inhibitors. Science. pii: eabb3405. https://doi.org/10.1126/science.abb3405

Chapter 4
Transmission Cycle of SARS-CoV and SARS-CoV-2

Tushar Yadav and Shailendra K. Saxena

Abstract Severe acute respiratory syndrome (SARS) is a pandemic that has shocked the world twice over the last two decades caused by a highly transmissible and pathogenic coronavirus (CoV). It causes disease in the lower respiratory tract in humans that was first reported in late 2002 in Guangdong province, China, and later on in December 2019 in Wuhan, China. The two viruses designated as SARS-CoV and SARS-CoV-2, respectively, originated probably from the bat and infected humans via carrier animals. The constant recombination and evolution in the CoV genome may have facilitated their cross-species transmission resulting in recurrent emergence as a pandemic. This chapter intends to accumulate recent findings related to CoV transmission and tentative molecular mechanisms governing the process.

Keywords SARS · CoV · Zoonotic · Transmission · Infection · Virus · Host

4.1 Introduction

Severe acute respiratory syndrome (SARS) is a high-risk viral disease usually characterized by fever, headache, and severe respiratory symptoms such as coughing, shortness of breath, and pneumonia (Peiris et al. 2004; Hu et al. 2017).

The original version of this chapter was revised. A correction to this chapter can be found at https://doi.org/10.1007/978-981-15-4814-7_17.

Tushar Yadav and Shailendra K. Saxena contributed equally as first author.

T. Yadav (✉)
Department of Zoology, Government Jawaharlal Nehru Smriti Postgraduate College, Shujalpur, Madhya Pradesh, India

S. K. Saxena
Centre for Advanced Research (CFAR)-Stem Cell/Cell Culture Unit, Faculty of Medicine, King George's Medical University (KGMU), Lucknow, India
e-mail: shailen@kgmcindia.edu

S. K. Saxena (ed.), *Coronavirus Disease 2019 (COVID-19)*, Medical Virology: from Pathogenesis to Disease Control, https://doi.org/10.1007/978-981-15-4814-7_4

These viruses can infect the respiratory, gastrointestinal, hepatic, and central nervous system of humans, livestock, birds, bat, mouse, and many other wild animals (Wang et al. 2006; Ge et al. 2013; Chen et al. 2020a, b). It first emerged in southern China in late 2002 and due to its fast transmission rate among humans, it immediately led to a global pandemic in 2003. More recently it has created havoc around the globe from December 2019 onwards, spreading over 100 countries and affecting thousands of people (Remuzzi and Remuzzi 2020). For this reason, it has been considered as a major public health threat in the twenty-first century (Zhong et al. 2003; Cui et al. 2019).

The causative agent of SARS is a coronavirus (CoV), more accurately SARS coronavirus (SARS-CoV), has been previously assigned to group 2b CoV, and is now a member of the lineage B of genus Betacoronavirus in the family Coronaviridae and subfamily Coronavirinae (Drexler et al. 2014; Chen et al. 2020b; Kumar et al. 2020). It shares similar genome organization with other coronaviruses, but exhibits a unique genomic structure which includes several specific accessory genes, including ORF3a, 3b, ORF6, ORF7a, 7b, ORF8a, 8b, and 9b (Hu et al. 2017). SARS coronavirus (SARS-CoV) uses angiotensin-converting enzyme 2 (ACE2) as a receptor and primarily infects ciliated bronchial epithelial cells and type II pneumocytes (Li et al. 2003; Qian et al. 2013).

A recent outbreak of a new CoV strain during December 2019 in China has drawn huge attention throughout the world. The administrative and scientific communities of China still are working towards the etiology, prevention and control, and drug development for the epidemic. On January 12th, 2020, the World Health Organization provisionally named the new virus as 2019 novel coronavirus (2019-nCoV) that later on renamed as severe acute respiratory syndrome coronavirus 2, SARS-CoV-2 (Gorbalenya et al. 2020). The continuous evolution and transformation of CoVs lead to sudden outbreaks to the different parts of the world suggesting that it may pose a serious global health hazard in a very short period. In today's changing climate and ecological balance, and the human-animal interactions, there is an increased risk of CoV disease outbreaks. This makes an utter requirement to focus on efficient measures to fight against CoVs.

A disease that usually occurs among animals but can infect humans in specific conditions is known as a zoonotic disease. They have largely affected the human population for the past hundreds of years. However, with passing time they have changed in several perspectives concerning their occurrences and pathogenicity (Rodriguez-Morales et al. 2020). As for now, several proofs indicate that CoV transmission occurs via "zoonotic spillover," a term indicating the transmission of a pathogen from a vertebrate animal to a human host. Although the mechanism of such transmission is not very clear and therefore it is a matter of concern, certain factors determine zoonotic spillovers such as behavioral characteristics of CoV and the susceptibility of a human host (Plowright et al. 2017).

The upcoming sections present a collection of recent findings on CoV infection. Additionally, the different factors responsible for the varied transmission mode of CoVs among animals and humans and the possible mechanism behind the process have been discussed.

4.2 Coronavirus Transmission Cycle

A coronavirus (SARS-CoV) is considered as the etiological agent of SARS. Further investigation proved that the first transmission of the virus to human hosts occurred probably in southern China in Guangdong province, from zoonotic reservoirs, including bats, Himalayan palm civets (*Paguma larvata*), and raccoon dogs (*Nyctereutes procyonoides*), the latter two of which are sold in exotic animal markets (Graham and Baric 2010).

Studies from the past suggest that SARS-CoV may also have a broad host range besides humans. SARS-CoV was transmitted directly to humans from market civets and is thought to have originated in bats (Cui et al. 2019).

Earlier, genetically similar CoVs were isolated from civet cats and raccoon dogs (Guan et al. 2003). Studies show that SARS-CoV has the ability to infect and produce disease in macaques and ferrets too, while did not produce any readily observable symptoms in cats (Fouchier et al. 2003; Martina et al. 2003). A recent study reports about 80% gene similarity between SARS-CoV-2 and SARS-CoV (Gralinski and Menachery 2020; Xu et al. 2020). Correspondingly, one more study reports a 96% sequence similarity between SARS-CoV-2 and the CoV isolated from *Rhinolophus affinis* indicating bats as virus source (Zhou et al. 2020). To date, there is not much clarity about SARS-CoV-2 host and it is reported to be snakes, minks, or other animals (Ji et al. 2020).

Figure 4.1 represents the tentative transmission path from a natural host to a human. The natural host of the CoV is considered as a bat (Li et al. 2020). While the species differ, CoV can still manage to migrate from its natural host to humans via intermediate host depending on its ability to access the host cell (Rodriguez-Morales et al. 2020). Since the last few decades, CoV has evolved to adapt to bind the receptors to enter inside the host's cells through its surface glycoproteins. These

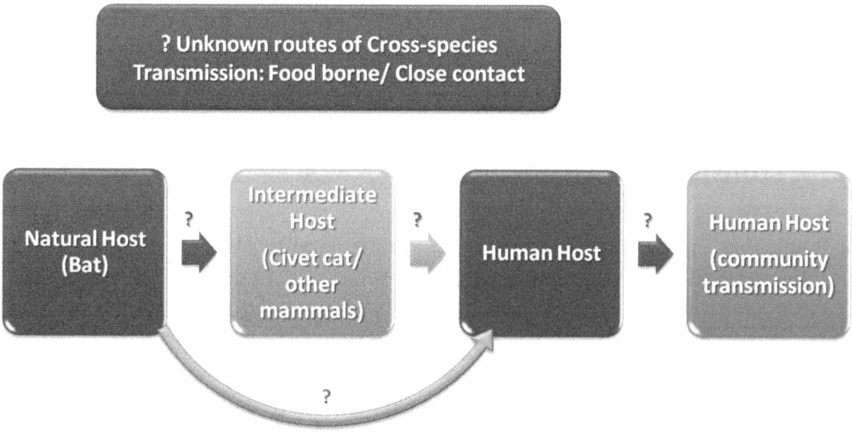

Fig. 4.1 The probable transmission path of SARS-CoV and SARS-CoV-2 from natural hosts to various hosts

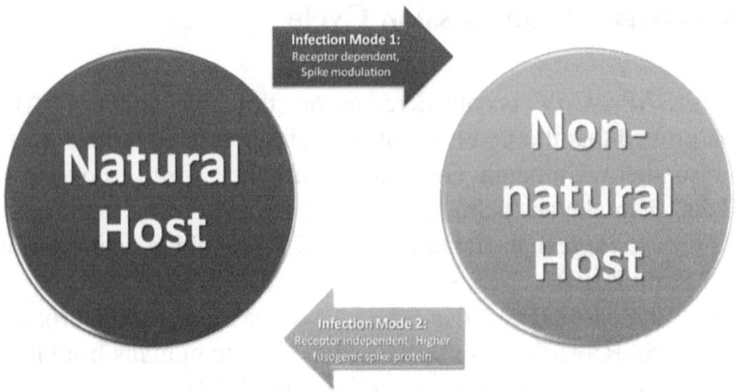

Fig. 4.2 Potential cross-species transmission mechanism from natural host to non-natural hosts (other animals and humans)

surface glycoproteins show significant variations that allow the virus to bind to varied mammalian host species (Rothan and Byrareddy 2020).

It has been known for decades that CoVs occasionally avoid the receptor-dependent entry into the host cell. The murine coronavirus strain JHM (MHV-JHM or MHV4) that codes for an extremely fusogenic spike protein may accomplish the infection via cell-to-cell spread mechanisms by giving up the known receptor-dependent entry route (Gallagher et al. 1992, 1993). In another study, MHV-JHM was found to infect CEACAM$^{-/-}$ (carcinoembryonic antigen-cell adhesion molecule) mice severely, a phenotype that was mapped specifically to the JHM strain spike protein (Miura et al. 2008). Based on these results, it can be assumed that there is an existence of a receptor-switching mechanism in CoVs leading to spike modularity and its tendency for recombination. Another speculation can be derived here about the higher fusogenic potential of the spike protein that minimizes its dependency on receptor-based cell entry (Nakagaki and Taguchi 2005; Graham and Baric 2010). The SARS-CoV and SARS-CoV-2 may have followed one of these mechanisms as depicted in Fig. 4.2.

4.3 Transmission Among Animals

In 2005, two individual research groups reported novel coronaviruses associated with human SARS-CoV, which were named SARS-CoV-related viruses or SARS-like coronaviruses, in horseshoe bats (genus *Rhinolophus*) (Lau et al. 2005; Li et al. 2005). Based on these studies it was understood that bats may have played a natural host for SARS-CoV while civets acted only as an intermediate. One more study exposed the coexistence of varied SARSr-CoVs in bat populations inhabiting one cave of Yunnan province, China, that was also the first information regarding human ACE2 (angiotensin-converting enzyme 2) as a receptor for bat SARS-like coronavirus (Ge et al. 2013; Hu et al. 2017). Further, it has already been known that the coronavirus genome frequently undergoes recombination (Lai and Cavanagh 1997), suggesting the high possibility of the emergence of new SARS-CoV through recombination of bat SARS-CoVs existing in same or another bat caves. Cui et al. (2019) speculated the production of SARS-CoV direct progenitor via recombination within bats, and thereafter it passed on to the farmed civets and other mammals leading to virus infection to civets by fecal-oral transmission. These virus-contained civets transported to the Guangdong market where they infected market civets and further mutated before affecting humans.

The phylogenetic investigation of novel CoVs suggests the existence of several cross-species transmission events; however, most of these events were transient spillover. The high recombination frequency of CoVs in bats suggests bats being a vital reservoir for CoV recombination and evolution (Banerjee et al. 2019).

4.4 Transmission from Animals to Human

The zoonotic origin of SARS-CoV-2 in Wuhan, China, can be strongly associated with the wet animal market since a large number of people who got an infection in the beginning were more or less exposed to it (Rothan and Byrareddy 2020). Several attempts were made to confirm the primary host or intermediary carriers from which the infection may have transmitted to humans.

Current research confirms more than 95% genomic similarity between SARS-CoV-2 and bat coronavirus, indicating bats as the most probable host of the former (Perlman 2020; Zhou et al. 2020). Besides bat, several other animal hosts were reported as a virus reservoir. Ji et al. (2020) demonstrated snakes as a possible virus reservoir for human infection while Lam et al. (2020) identified SARS-CoV-2-related coronaviruses in pangolins (*Manis javanica*). Stated the possibility of minks being intermediate hosts for SARS-CoV-2.

One more remarkable phenomenon observed for SARS-CoV was that the human strain recovered at the time of epidemic retained efficient hACE2/cACE2 recognition; however, the in vitro adapted civet strains quickly achieved hACE2 recognition (Sheahan et al. 2008). These data indicate the competent human/civet ACE2 recognition as a key factor to support SARS-CoV in human populations, offering an animal reservoir for continual persistence.

The main culprit for SARS-CoV and SARS-CoV-2 in humans is considered as bats since they are known to contain a wide variety of coronaviruses, although the mechanism for virus zoonotic spillover is still unclear. The pieces of evidences suggest the occurrence of recombination events among SARS-CoVs exist in the neighbouring bat population. Such phenomena may be responsible for the series of recombination within the S gene and around ORF8 leading to the origin of SARS-CoV direct progenitor. Moreover, it is expected that from here the spillover took place from bats to civets and later on to the people residing near the location or due to indulgence in wildlife trade of infected animals (Hu et al. 2017; Ahmad et al. 2020; Lu et al. 2020).

The spillover proceeds via several consecutive events that facilitate CoVs to establish infection in humans. The probability of animal-to-human transmission is ruled by various factors such as the dynamics of disease in an animal host, level of virus exposure, and the susceptibility of human population. All these factors can be summarized into three major stages that depict the way of virus transmission. The primary stage defines the pathogen pressure on human host, i.e., the amount of virus interacting with humans at a particular instant regulated by virus prevalence and dispersal from the animal host, followed by its survival, development, and distribution outside the animal host. In the next stage, the behavior of humans and vector defines the chances of viral exposure, the route of entry, and the dose of the virus. The last stage is influenced by genetics, the physiological and immunological status of the human host along with stage two factors determining the possibility and severity of infection (Plowright et al. 2017). The aforementioned stages create a barrier for transmission of the virus to the next level, and spillover necessitates the virus to surpass all barriers to establish an infection in the upcoming host.

4.5 Transmission Among Humans

Data from various studies so far implicate the zoonotic origin of SARS-CoV and SARS-CoV-2, and its fast spread among humans confirms person to person transmission. Many research works present added information on such modes of transmission. SARS being an airborne virus, transmit via the same way as cold and flu do. The virus spreads by an infected person on coughing or sneezing leaving small droplets in the air or by stool. So the person who inhales such droplets or touches the infected surfaces may also get infected.

In recent works, live SARS-CoV-2 has been detected in the stool of patients evidencing the subsistence of SARS-CoV in the gastrointestinal tract justifying the gastrointestinal symptoms, probable recurrence, and transmission of the virus via fecal-oral route (Gu et al. 2020; Holshue et al. 2020). However, it is not sure whether the consumption of virus-contaminated food may cause infection and transmission (Wu et al. 2020).

Ghinai et al. (2020) found that person-to-person transmission of SARS-CoV-2 may occur due to prolonged and unprotected exposure with the infected person suggesting constant pathogen pressure leading to infection and disease.

A case of SARS-CoV-2 transmission along four successive generations has been studied. Such incidence produces an example of sustained human to human transmission (Phelan et al. 2020; WHO 2020). So far the SARS-CoV-2-infected person acted as a major infection source and respiratory droplets as the main route of transmission, along with aerial droplets and close contact (Jin et al. 2020).

The virus infection commences via binding specific host receptors then fusing with the cell membrane. Reports state that the receptor-binding domain (RBD) of virus spikes binds with ACE2 receptor of the potential host cell in case of SARS-CoV human-to-human transmission (Jaimes et al. 2020; Wan et al. 2020). The most interesting feature is that SARS-CoV-2 and SARS-CoV spikes share RBD sequence similarity strongly suggesting their common route of entry into the host cells via the ACE2 receptor (Wan et al. 2020). Currently, there is inadequate information on the transmission of SARS-CoV and SARS-CoV-2 from pet animals like dog and cat to the human; however in a fast-evolving situation it is difficult to predict the future.

Executive Summary

- The recurrent emergence of pathogenic CoVs indicates the disturbances in their ecological niche.
- There is still no clarity on the potential animal reservoir for the virus.
- The alleged zoonotic origin and cross-species transmission of SARS-CoV and SARS-CoV-2 need further attention to confirm the underlying adaptation-evolution mechanism.
- Understanding the molecular basis for ACE2 receptor usage by different SARS-CoV strains is vital to obtain clarity on cross-species transmission and check on possible future disease outbreaks.

4.6 Conclusions

Numerous viruses have existed in nature for years without affecting the human population, and their recurrent spillover on other animals and humans is the consequence of man-made activities. Therefore, the best way to keep them away is to keep a barrier between their natural reservoirs and civilization.

4.7 Future Perspectives

There is a need to scrutinize more animal models for infection and transmission of SARS among animals and humans. We need to study the effect of ethnic and cultural differences on CoV transmission and pathogenesis.

References

Ahmad T, Khan M, Haroon, Musa TH, Nasir S, Hui J, Bonilla-Aldana DK, Rodriguez-Morales AJ (2020) COVID-19: zoonotic aspects. Travel Med Infect Dis 101607. https://doi.org/10.1016/j.tmaid.2020.101607

Banerjee A, Kulcsar K, Misra V, Frieman M, Mossman K (2019) Bats and coronaviruses. Viruses 11(1):41. https://doi.org/10.3390/v11010041

Chen H, Guo J, Wang C, Luo F, Yu X, Zhang W, Li J, Zhao D, Xu D, Gong Q, Liao J, Yang H, Hou W, Zhang Y (2020a) Clinical characteristics and intrauterine vertical transmission potential of COVID-19 infection in nine pregnant women: a retrospective review of medical records. Lancet 395:809–815. https://doi.org/10.1016/S0140-6736(20)30360-3

Chen Y, Liu Q, Guo D (2020b) Emerging coronaviruses: genome structure, replication, and pathogenesis. J Med Virol 92:418–423. https://doi.org/10.1002/jmv.25681

Cui J, Li F, Shi ZL (2019) Origin and evolution of pathogenic coronaviruses. Nat Rev Microbiol 17 (3):181–192

Drexler JF, Corman VM, Drosten C (2014) Ecology, evolution and classification of bat coronaviruses in the aftermath of SARS. Antiviral Res 101:45–56. https://doi.org/10.1016/j.antiviral.2013.10.013. PMID: 24184128

Fouchier RA, Kuiken T, Schutten M, van Amerongen G, van Doornum GJ, van den Hoogen BG, Peiris M, Lim W, Stohr K, Osterhaus AD (2003) Aetiology: Koch's postulates fulfilled for SARS virus. Nature 423(6937):240. https://doi.org/10.1038/423240a

Gallagher TM, Buchmeier MJ, Perlman S (1992) Cell receptor independent infection by a neurotropic murine coronavirus. Virology 191:517–522

Gallagher TM, Buchmeier MJ, Perlman S (1993) Dissemination of MHV4 (strain JHM) infection does not require specific coronavirus receptors. Adv Exp Med Biol 342:279–284

Ge X, Li J, Yang X et al (2013) Isolation and characterization of a bat SARS-like coronavirus that uses the ACE2 receptor. Nature 503:535–538. https://doi.org/10.1038/nature12711

Ghinai I et al (2020) First known person-to-person transmission of severe acute respiratory syndrome coronavirus 2 (SARS-CoV-2) in the USA. Lancet. https://doi.org/10.1016/S0140-6736(20)30607-3

Gorbalenya AE, Baker SC, Baric RS et al (2020) The species severe acute respiratory syndrome related coronavirus: classifying 2019-nCoV and naming it SARS-CoV-2. Nat Microbiol 5:536–544. https://doi.org/10.1038/s41564-020-0695-z

Graham RL, Baric RS (2010) Recombination, reservoirs, and the modular spike: mechanisms of coronavirus cross-species transmission. J Virol 84(7):3134–3146. https://doi.org/10.1128/JVI.01394-09

Gralinski LE, Menachery VD (2020) Return of the coronavirus: 2019-nCoV. Viruses 12(2):135. https://doi.org/10.3390/v12020135

Gu J, Han B, Wang J (2020) COVID-19: gastrointestinal manifestations and potential fecal-oral transmission. Gastroenterology (Article in Press). https://doi.org/10.1053/j.gastro.2020.02.054

Guan Y, Zheng BJ, He YQ et al (2003) Isolation and characterization of viruses related to the SARS coronavirus from animals in southern China. Science 302(5643):276–278

Holshue ML, DeBolt C, Lindquist S et al (2020) First case of 2019 novel coronavirus in the United States. N Engl J Med 382:929–936. https://doi.org/10.1056/NEJMoa2001191

Hu B, Zeng LP, Yang XL et al (2017) Discovery of a rich gene pool of bat SARS-related coronaviruses provides new insights into the origin of SARS coronavirus. PLoS Pathog 13 (11):e1006698. https://doi.org/10.1371/journal.ppat.1006698

Jaimes JA, Millet JK, Stout AE, Andre NM, Whittaker GR (2020) A tale of two viruses: the distinct spike glycoproteins of feline coronaviruses. Viruses 12(1):83. https://doi.org/10.3390/v12010083

Ji W, Wang W, Zhao X, Zai J, Li X (2020) Homologous recombination within the spike glycoprotein of the newly identified coronavirus may boost cross-species transmission from snake to human. J Med Virol 92:433–440

Jin YH, Cai L, Cheng ZS et al (2020) A rapid advice guideline for the diagnosis and treatment of 2019 novel coronavirus (2019-nCoV) infected pneumonia (standard version). Mil Med Res 7:4. https://doi.org/10.1186/s40779-020-0233-6

Kumar S, Maurya VK, Prasad AK, Bhatt MLB, Saxena SK (2020) Structural, glycosylation and antigenic variation between 2019 novel coronavirus (2019-nCoV) and SARS coronavirus (SARS-CoV). VirusDis 31:13–21. https://doi.org/10.1007/s13337-020-00571-5

Lai MM, Cavanagh D (1997) The molecular biology of coronaviruses. Adv Virus Res 48:1–100

Lam TTY, Shum MHH, Zhu HC et al (2020) Identification of 2019-nCoV related coronaviruses in Malayan pangolins in southern China. Nature. https://doi.org/10.1038/s41586-020-2169-0. https://doi.org/10.1101/2020.02.13.945485

Lau SKP, Woo PCY, Li KSM et al (2005) Severe acute respiratory syndrome coronavirus-like virus in Chinese horseshoe bats. Proc Natl Acad Sci U S A 102:14040–14045

Li W, Moore MJ, Vasilieva N et al (2003) Angiotensin-converting enzyme 2 is a functional receptor for the SARS coronavirus. Nature 426:450–454

Li W, Shi Z, Yu M et al (2005) Bats are natural reservoirs of SARS-like coronaviruses. Science 310 (5748):676–679

Li B, Si HR, Zhu Y, Yang XL, Anderson DE, Shi ZL, Wang LF, Zhou P (2020) Discovery of bat coronaviruses through surveillance and probe capture-based next-generation sequencing. mSphere 5:e00807–e00819. https://doi.org/10.1128/mSphere.00807-19

Lu R, Zhao X, Li J et al (2020) Genomic characterisation and epidemiology of 2019 novel coronavirus: implications for virus origins and receptor binding. Lancet 395(10224):565–574. https://doi.org/10.1016/S0140-6736(20)30251-8

Martina BE, Haagmans BL, Kuiken T et al (2003) Virology: SARS virus infection of cats and ferrets. Nature 425(6961):915

Miura TA, Travanty EA, Oko L, Bielefeldt-Ohmann H, Weiss SR, Beauchemin N, Holmes KV (2008) The spike glycoprotein of murine coronavirus MHV-JHM mediates receptor-independent infection and spread in the central nervous systems of Ceacam1a$^{-/-}$ mice. J Virol 82:755–763

Nakagaki K, Taguchi F (2005) Receptor-independent spread of a highly neurotropic murine coronavirus JHMV strain from initially infected microglial cells in mixed neural cultures. J Virol 79:6102–6110

Peiris JS, Guan Y, Yuen KY (2004) Severe acute respiratory syndrome. Nat Med 10:S88–S97. https://doi.org/10.1038/nm1143. PMID: 15577937

Perlman S (2020) Another decade, another coronavirus. N Engl J Med 382:760–762

Phelan AL, Katz R, Gostin L (2020) The novel coronavirus originating in Wuhan, China: challenges for global health governance. JAMA 323(8):709–710. https://doi.org/10.1001/jama.2020.1097

Plowright RK, Parrish CR, McCallum H et al (2017) Pathways to zoonotic spillover. Nat Rev Microbiol 15(8):502–510. https://doi.org/10.1038/nrmicro.2017.45

Qian Z, Travanty EA, Oko L et al (2013) Innate immune response of human alveolar type II cells infected with severe acute respiratory syndrome-coronavirus. Am J Respir Cell Mol Biol 48:742–748

Remuzzi A, Remuzzi G (2020) COVID-19 and Italy: what next? Lancet 395(10231):1225–1228. https://doi.org/10.1016/S0140-6736(20)30627-9

Rodriguez-Morales AJ, Bonilla-Aldana DK, Balbin-Ramon GJ, Rabaan AA, Sah R, Paniz-Mondolfi A, Pagliano P, Esposito S (2020) History is repeating itself: probable zoonotic spillover as the cause of the 2019 novel coronavirus epidemic. Infez Med 28(1):3–5

Rothan HA, Byrareddy SN (2020) The epidemiology and pathogenesis of coronavirus disease (COVID-19) outbreak. J Autoimmun 26:102433. https://doi.org/10.1016/j.jaut.2020.102433

Sheahan T, Rockx B, Donaldson E, Corti D, Baric R (2008) Pathways of cross-species transmission of synthetically reconstructed zoonotic severe acute respiratory syndrome coronavirus. J Virol 82(17):8721–8732. https://doi.org/10.1128/JVI.00818-08

Wan Y, Shang J, Graham R, Baric RS, Li F (2020) Receptor recognition by novel coronavirus from Wuhan: an analysis based on decade-long structural studies of SARS. J Virol 94. https://doi.org/10.1128/JVI.00127-20

Wang LF, Shi Z, Zhang S, Field H, Daszak P, Eaton BT (2006) Review of bats and SARS. Emerg Infect Dis 12(12):1834–1840. https://doi.org/10.3201/eid1212.060401

WHO (2020) How does COVID-19 spread? https://www.who.int/news-room/q-a-detail/q-a-coronaviruses. Accessed 16 Mar 2020

Wu D, Wu T, Liu Q, Yang Z (2020) The SARS-CoV-2 outbreak: what we know. Int J Infect Dis (Article in Press). https://doi.org/10.1016/j.ijid.2020.03.004

Xu J, Zhao S, Teng T et al (2020) Systematic comparison of two animal-to-human transmitted human coronaviruses: SARS-CoV-2 and SARS-CoV. Viruses 12:244. https://doi.org/10.3390/v12020244

Zhong NS, Zheng BJ, Li YM, Poon, Xie ZH, Chan KH et al (2003) Epidemiology and cause of severe acute respiratory syndrome (SARS) in Guangdong, People's Republic of China, in February, 2003. Lancet 362(9393):1353–1358. PMID: 14585636

Zhou P, Yang X, Wang X et al (2020) A pneumonia outbreak associated with a new coronavirus of probable bat origin. Nature 579:270–273. https://doi.org/10.1038/s41586-020-2012-7

Chapter 5
Host Immune Response and Immunobiology of Human SARS-CoV-2 Infection

Swatantra Kumar, Rajni Nyodu, Vimal K. Maurya, and Shailendra K. Saxena

Abstract One of the most serious viral outbreaks of the decade, infecting humans, originated from the city of Wuhan, China, by the end of December 2019, has left the world shaken up. It is the successor infection of severe acute respiratory syndrome coronavirus (SARS-CoV) named as SARS-CoV-2 causing a disease called as COVID-19 (Coronavirus disease-19). Being one of the most severe diseases in terms of transmission, this disease agitates the immune system of an individual quite disturbingly which at times leads to death, which is why it has become the need of the hour to step forward to extensively involve in understanding the genetics, pathogenesis, and immunopathology of SARS-CoV-2 in order to design drugs to treat or to design a vaccine to prevent. In this chapter, we have tried to review and summarize the studies done so far to understand the host–pathogen relationship and the host immune response during COVID-19 infection. One of the recent developments regarding the understanding of SARS-CoV-2 infection is the mechanism of immune evasion involved during the pathogenesis and cytokine storm syndrome during infection in the patient against which a drug called as Hydroxychloroquine has been designed. Comprehensively, we have tried to give an immunological insight into the SARS-CoV-2 infection in order to understand the possible outcome for any therapeutic advancement.

Keywords Cytokine storm syndrome · Immunopathogenesis · ARDS · T cell response · B cell response · IFN-mediated signaling

---

S. Kumar · R. Nyodu · V. K. Maurya
Centre for Advanced Research (CFAR), Faculty of Medicine, King George's Medical University (KGMU), Lucknow, India

S. K. Saxena (✉)
Centre for Advanced Research (CFAR)-Stem Cell/Cell Culture Unit, Faculty of Medicine, King George's Medical University (KGMU), Lucknow, India
e-mail: shailen@kgmcindia.edu

S. K. Saxena (ed.), *Coronavirus Disease 2019 (COVID-19)*, Medical Virology: from Pathogenesis to Disease Control, https://doi.org/10.1007/978-981-15-4814-7_5

5.1 Introduction

With such an avalanche of reported cases of CoVID-19 (Coronavirus disease-19) from all across the globe, the family of Coronavirus has pretty nearly brought the entire human race under its feet since late December 2019 (Khan et al. 2020). "Corona" loosely meaning "halo" or "crown" in Latin refers to the structure seen by the capsid and RNA. "Coronavirus" was actually termed during the imaging of the viral family Coronaviridae, due to the circular shape of the virus itself (Li 2016). CoVID-19 is caused by the novel strain of Coronavirus named as SARS-CoV-2 due to its homology with SARS infection and has been declared as a pandemic by the WHO (World Health Organization) on March 2020 (World Health Organization 2020). This novel strain of CoV debuted into the human host causing severe pneumonia and respiratory disorders (Wang et al. 2020). SARS-CoV-2 has managed to rattle the host immune system very successfully. Reports suggest that, analogous to its predecessor (SARS-CoV), who belongs to the same genera and family, the origin of SARS-CoV-2 has also been reported to be from the species of bats (Li et al. 2020). This strain of coronavirus is highly contagious, which is the root cause of CoVID-19 being spread rapidly and causing higher number of casualties ever since the first case has been reported.

5.2 Family of Coronaviruses

Coronaviruses (CoVs) are a group of viruses that belong to the family *Coronaviridae*, infecting humans along with other species (not every CoVs), and are respiratory illness causing viruses. The first encounter with the CoV was seen in 1960s, which was named as HCoV-OC43 and HCoVs 229E (Drosten et al. 2003). Until late December 2019, six such groups of CoVs were being known: HCoV-OC43, HCoV-HKU1, HCoV-229E, SARS-CoV, HCoV-NL63, and MERS-CoV are among them. CoVs are categorized under four genera: Alpha CoVs (HCoV-NL63, HCoV-229), Beta CoVs (HCoV-OC43, SARS-CoV, HCoV-HKU1, and MERS-CoV), Gamma CoVs, and Delta CoVs (Fehr and Perlman 2015).

5.3 Entry of CoV into Host Cells

Coronaviruses uses a very spiky-shaped protein, S protein, to infect a cell by binding to the membrane of the cell. COVID-19 (SARS-CoV-2) and SARS-CoV share a receptor-binding unit whose domain structure is similar, suggesting that COVID-19 (SARS-CoV-2) uses ACE2 receptor in humans for infection (Yan et al. 2020). The spike protein binds to this ACE2 receptor on the host cell surface and gets pinched inside the host cell. Studies has been conducted which shows the role of an enzyme

Furin present in the host cells, plays a crucial role in SARS-CoV-2 entry, and can be a distinguishing feature defining the severity of SARS-CoV-2, since it is absent in SARS-CoV (Walls et al. 2020). This enzyme activates SARS-CoV-2 whereas the SARS-CoV and MERS-CoV during entrance into the host cell do not encounter this activated site. Since furin is expressed in various human organs such as the lungs, small intestine, and liver, the infection in human has been seen to be very vigorous and can be predicted to be potentially infecting multiple human organs. This site could possibly affect the transmission as well as the stability of the virus.

5.4 Antigen Presentation During Human SARS-CoV-2 Infection

As an antiviral mechanism, antigen Presenting Cells (APC) are involve in the presentation of viral antigenic peptides in complexed with MHC (major histocompatibility complex) class I and class II molecules to CD8 and CD4 T cells. The selection of peptides and presentation technique of the host leads to a better understanding of cellular immunity and vaccine advancement. During any viral infection, DCs (dendritic cells) play a very important role as an APC. DCs are a linkage between innate and adaptive immunity. Studies deciphering the mechanism of antigen presentation during SARS-CoV-2 infection is not studied well therefore, the mechanism of antigen presentation can be understood based on the available data of predecessor strain infection (SARS-CoV & MERS-CoV) due to its analogy (Chen et al. 2010). Because DCs are found in the respiratory tract and react back whenever there is an inflammation response, DCs are found to be a potential candidate in antigen presentation during SARS infection and also in understanding the immunopathology of SARS (Lau et al. 2012).

During SARS infection, the upregulation of few chemokines such as IP-10 and MP1 is seen very significantly, also few of the antiviral cytokines are found to be low in expression such as IFN-alpha, IFN-beta, and IFN-gamma, and TNF-alpha and IL-6 are found to be at moderate upregulation (Kuri and Weber 2010). Modulation of Toll-like receptors from TLR-1 to TLR-10 was seen to be at the same level; hence no modulation but chemokine receptors such as CCR5, CCR3, and CCR1 are found to be at significant level of upregulation (Law et al. 2009). When similar type of study is conducted in patients infected with Middle East Respiratory Syndrome Coronavirus (MERS-CoV), it has been observed that this virus infects DCs very prolifically by inducing higher expression of IFN-gamma and even cytokines and chemokine related with IFN-gamma are found to be at a higher level. Altogether, antigen presentation in case of MERS-CoV-infected dendritic cells is seen to be significantly higher than in SARS-CoV-infected dendritic cells.

In case of SARS-CoV, the antigen presentation is done most importantly by MHC-I followed by MHC-II (Wieczorek et al. 2017). Studies were done in human macrophages during SARS-CoV infection indicating an interesting observation that severe acute respiratory syndrome coronavirus infects human macrophages due to antibody enhancement (ADE) mediated by IgG. However, macrophages infected with SARS-CoV did not show productive viral replication (Yip et al. 2014). MHC class II and I are further analyzed for epitope mapping which is one of the key steps towards vaccine development and which is studied intensively in case of SARS-CoV-2 infection (Sarkar et al. 2020). Post MHC-I analysis, it was observed that the protein sequence of CoVID-19 is introduced by alleles of HLA-C rather than HLA-A and HLA-B (Prompetchara et al. 2020). Alleles of HLA are studied in case of SARS-CoV infection and have been categorized into protection alleles (HLA-Cw1502, HLA-DR0301, and HLA-A*0201) and susceptible alleles (HLA-B*0703, HLA-Cw*0801, HLA-DR B1*1202, and HLA-B*4601) (Wang et al. 2011). In case of MERS-CoV infection, molecules of MHC class II, HLA-DQB1*02:0, and HLA-DRB1*11:01 are related with the susceptibility to infection (Fast and Chen 2020; Hajeer et al. 2016). While analyzing the viral genome of COVID-19 for epitope prediction, around 405 T cell epitopes have been determined which exhibits effective affinity towards MHC class I and II in addition two potent neutralizing epitopes (B cell epitope) based on Spike (S) protein were found (Fast and Chen 2020).

5.5 Humoral Immune Response

Humoral immune response is an antibody-mediated immune response. T helper cells assist B cells to differentiate into plasma cells, which in return produces antibodies (Abs) specific to a viral antigen (Ag). In order to limit infection, an antibody which is of neutralizing nature is efficient in fully blocking the virus from entering into host cells and hence plays a very intense protective role at a later stage of infection and also prevents relapse of infection in the future. In case of SARS-CoV, in order to enhance the humoral immune response, both B and T cell epitopes are being studied extensively and mapped for their structural as well as the envelope protein (S, N, M, E) (Channappanavar et al. 2014). During the infection of SARS-CoV-2 it is studied and known that ACE2 (Angiotensin-Converting Enzyme 2) is the receptor present in the host cells and hence the site (B cell epitope site) wherein the Ab-binding site is located is a critical feature to be studied for a valuable insight (Yu et al. 2007). It has been observed that SARS-CoV and 2019-nCoV spike proteins are similar in their major structural and T cell epitope (Kumar et al. 2020).

In order to produce neutralizing Abs, both the proteins need to have B cell epitopes as their receptor binding domains (RBD). A very efficient T cell response has been found to be in relation with the production of higher neutralizing antibodies. T cell epitopes do not require a specific location in contrast to B cell epitopes; hence, it is located anywhere in viral protein. The helper T cell plays a role in isotype switching also and in case of SARS-CoV, the antibody profile of this virus produces IgM and IgG and at a later phase sero-conversion has been observed which is mediated by the helper T cells (Li et al. 2008). IgM disappears at the end of week 12 whereas IgG has been found to last for a longer time pointing out to the probability of IgG being a potent protector Ab during the infection (Li et al. 2003). Current evidence strongly indicates that Th1 type response is key to the successful control of SARS-CoV and MERS-CoV and probably true for SARS-CoV-2 as well (Yong et al. 2019).

5.6 Cellular Immune Response

Cellular immune response is a mechanism of adaptive immunity. Cellular immunity in contrast to the humoral immune response can be seen inside the infected cells, which is mediated by T-lymphocytes. Helper T cells direct the overall adaptive immune response while cytotoxic T cells play a vital role in clearance and killing of viral infected cells. For any effective vaccine advancement, cellular immunity provided by T cells is very much essential as shown by the mouse model experiment on MERS-CoV and SARS-CoV (Yong et al. 2019) wherein their reports suggested that the lack of T cells resulted in no viral clearance in infected mice, hence explaining the importance of T cells in viral infection (Lee et al. 2012).

Referring back to the case of infection caused by Severe Acute Respiratory Syndrome Coronavirus and Middle East Respiratory Syndrome Coronavirus, it is reported that $CD4^+$ (TNFα, IL-2, and IFN) and $CD8^+$ (TNFα, IFNγ) memory T cells could persist in SARS-CoV-recovered patient for 4 years and can function by proliferating T cell, producing IFN-gamma, and by DTH response (Kuri and Weber 2010). When investigated 14 out of 23 SARS-recovered patients, post 6 years of infection, it was reported that distinct T cell memory responded to the S library of peptide of SARS-CoV (Channappanavar et al. 2014). Similar findings of distinct $CD8^+$ T cells were seen during the case of MERS-CoV clearance in a mouse model too (Coleman et al. 2016). Hence, this information can be useful in case of SARS-CoV-2 as well. However, in case of SARS-CoV-2 recent reports suggest that the PBMCs of SARS-CoV-2 infected individuals have shown efficient reduction in the $CD8^+$ and $CD4^+$ T cell counts, which may results in compromised T memory cell generation and persistence in SARS-CoV-2 survivors.

5.7 Cytokine Storm Syndrome in Patients Infected with SARS-CoV-2

Cytokine storm syndrome is when there is a fatal blow up of cytokines due to over-reaction by the human body's immune system in response to an intruder. The *Lancet* has published a report on cytokine storm syndrome as being one of the causes of CoVID-19 severity (Coleman et al. 2016). This report is backed up by the data on one of the major factors of death due to SARS-CoV-2 infection, which is acute respiratory distress syndrome (ARDS). ARDS has a significant relation with cytokine storm syndrome because, during ARDS, the immune effector cells have been shown to release huge amounts of chemokines and proinflammatory cytokines, which result in a fatal unconfined or uncontrolled systemic inflammatory response (Yao et al. 2020). Previous pandemics caused by coronaviruses such as MERS-CoV and SARS-CoV have also shown such a massive release of chemokines and cytokines: in case of SARS-CoV, CCL2, CCL3, CCl5, CXCL8, CXCL9, CXCL10, etc. and IL-12, IL-18, IL-6, IL-1beta, IL-33, IFN-alpha, IFN-gamma, TNF- alpha & TGF-beta. In case of MERS-CoV, the elevation was seen in the levels of cytokines-IFN-α, IL-6, and chemokine such as CXCL-10, CCL-5, and CXCL-8 (Zheng et al. 2020).

ARDS due to cytokine storm triggers a damaging attack to the body by the immune system causing failure of multiple organs subsequently and leads to death as reported in the case of SARS-CoV-2 outbreak which was the same as the cases of SARS-CoV and MERS-CoV infection reported previously. Lately, drugs targeting IL-18, IL-1, IL-6, and Interferon-gamma have been found effective in treating cytokine storm syndrome in other viral infections for the treatment and therefore may be used for the treatment of the COVID-19 patients for reducing the severity (Cameron et al. 2007). However, one of such drugs falling under the same category, which blocks IL-6, has been reported to be efficient in a few cases of COVID-19 in China (Mehta et al. 2020).

5.8 Immune Evasion Strategies for Coronaviruses

Human CoVs are the one of the most pathogenic viral infections that develops various immune evasion strategies. Studies have come up with reports supporting the fact that the family of CoVs are significantly able to suppress human immune responses by evading the immune detection mode smartly (Kikkert 2020). This immune evasion property might explain longer incubation period, which is of 2–11 days, moderately, if not completely. Immune evasion helps them to efficiently dodge the detection by cellular PRRs of host immune response at the initial phase of infection. The three recent CoVs such as SARS-CoV, MERS-CoV, and SARS-CoV-2 share the same component of immune evasion method since all of them belong to the same genera, Betacoronavirus. The strategy of how these CoVs evades and regulates human immune responses has been a highly talked, studied, and evaluated topic for a very long time. Several studies have been carried out on SARS-CoV and MERS-CoV, which can also be referred in the case of SARS-CoV-2. During SARS-CoV infection, the isolation of viral dsRNA takes place inside the double membrane vesicles (DMVs), which is the probable shield of viral PAMPs from detection by cytosolic PRRs.

In order to exist and expand inside a host with inbuilt strong antiviral IFN immune responses, CoVs have been seen to employ different levels of strategies (Fig. 5.1) against the innate immune responses, especially type I IFN responses starting with the IFN signaling, induction of IFN, or antiviral action of ISG products

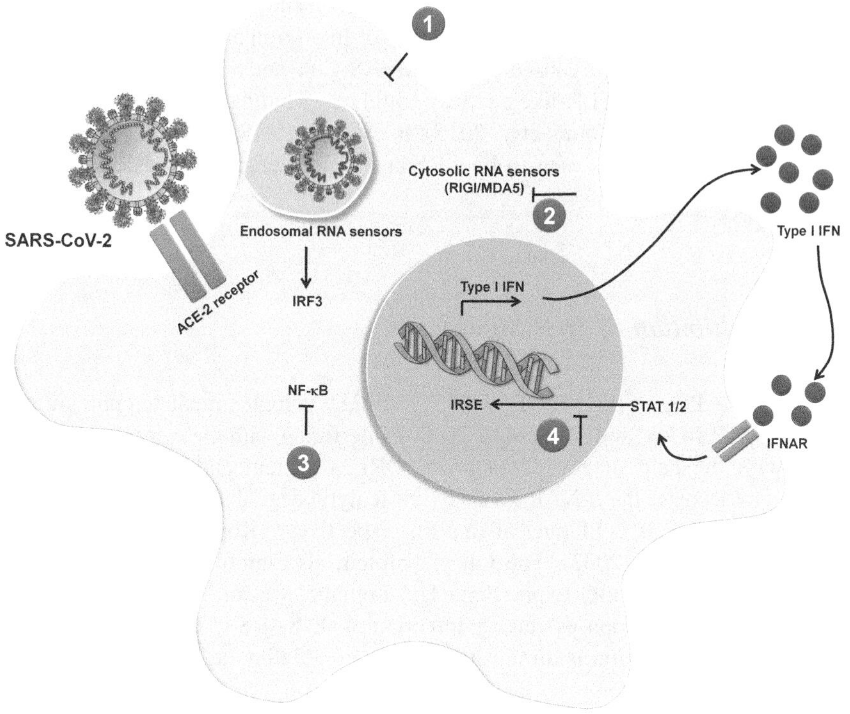

Fig. 5.1 Immune escape mechanism by SARS-CoV. Coronavirus interferes with multiple steps during initial innate immune response including RNA sensing (1, 2), signaling pathways of type 1 IFN production (3), STAT ½ activation downstream of IFN/IFNAR (4)

(Channappanavar et al. 2016). In any viral infection, interferons play the role of a very potent cytokine in order to control the infection. This suggests that CoVs play its smart game by interfering with the core system of IFN or it can also do so by destroying the key regulators.

5.9 Immune Escape Strategies

5.9.1 Inhibition of IFN Induction

To understand this strategy briefly, both the CoVs, SARS-CoV and MERS-CoV, seem to be less generous regarding inducing IFN into most of the cell types. This less or delayed induction of IFN activates the proinflammatory cytokines and macrophages into the lungs and hence results in leakage of vascular vessels and also impairs the adaptive immune responses (Channappanavar et al. 2016). Exceptional cell type wherein the IFN induction is at a relatively higher level is pDCs, which have shown to be expressing increased levels of Interferon alpha or beta (IFN-alpha & -beta) in case of disease caused by both MERS-CoV and SARS-CoV. This occurs through TLR7. Hence, TLR-like pathway could be a candidate for more core studies in case of SARS-CoV (Totura et al. 2015). In case of MERS-CoV, the protein which inhibits IFN induction is seen to be ORF4a which interacts with dsRNA and the cofactor of RLR, PACT.

5.9.2 Inhibition of IFN Signaling

From IFN docking to ISGF3, STAT1/STAT2/IRF9 complex mediated upregulation of ISGs, SARS-CoV and MERS-CoV act by interfering with the signal transduction chain with the help of several proteins. ORF3a protein and ORF6 protein are reported to decrease the IFNAR levels by proteolytic degradation and ubiquitination and by disrupting STAT1 nuclear import, respectively (Kopecky-Bromberg et al. 2007; Frieman et al. 2007). The ORF4a protein also inhibits IFN induction and hence moderates the ISG expression. The immune evasion strategy also evades adaptive immune responses during infection with SARS-CoV and MERS-CoV wherein T cell activation is diminished by downregulating antigen presentation by MHC I and II molecules.

5.10 Conclusions

The emergence and outbreak caused by SARS-CoV-2 is determined by the disruption of the host immune system by the virus. The virus strain has been observed to disturb the immune system by evasion of the immune response when a person is infected. It is on the safer side to predict that coronaviruses, having the biggest RNA genome so far, do not depend upon sole harmfulness factors; however they utilize a few layers of hostile to IFN procedures such as factors which include the type of virus, the titer of virus, and the load of virus. Else they would not have the strategy to exist, and even expand or mutate to a newer territory with strong antiviral IFN reactions (host). With just their genetic blueprint made available so far, it has become difficult for all research groups to identify the potential molecular targets of the virus. Immunopathology studies of disease caused by SARS-CoV-2 are still being investigated globally. However, with reference to the studies done so far on its predecessor infection, it can be concluded that few of the mechanisms which explain the severity of the disease caused by SARS-CoV-2 is the enzyme Furin, which is found at its activation site, and the mechanism of cytokine storm and immune evasion.

5.11 Future Perspectives

There are many hurdles in designing a vaccine against CoVID-19 due to the fact that different age group and individuals with pre-existing conditions react differently to this disease because of their differences in immune responses, which is why one-size vaccine will not be enough but still research groups are indulged in screening vaccine and the current vaccine candidate used is an antigen of the spike protein of coronavirus. Due to the lack of many failed antiviral strategies in order to efficiently treat infections by coronavirus, scientists are trying to come up with preventive measure such as vaccination. Taking into account the previous cases of coronavirus infection such as the infection of SARS-CoV and MERS-CoV, research groups have managed to come up till the stage of clinical trials of few vaccines and which is an advancement when it comes to the fight against such outbreaks. Hence, it will pave the way for some advancement in designing drugs and vaccine against CoVID-19 too. This is possibly a positive outcome for the vaccine candidate for SARS-CoV-2 disease in the future.

Acknowledgments and Disclosures The authors are grateful to the Vice Chancellor, King George's Medical University (KGMU), Lucknow, India, for encouraging this work. The authors have no other relevant affiliations or financial involvement with any organization or entity with a financial interest in or financial conflict with the subject matter or materials discussed in the manuscript apart from those disclosed.

References

Cameron MJ, Ran L, Xu L, Danesh A, Bermejo-Martin JF, Cameron CM, Muller MP, Gold WL, Richardson SE, Poutanen SM, Willey BM, DeVries ME, Fang Y, Seneviratne C, Bosinger SE, Persad D, Wilkinson P, Greller LD, Somogyi R, Humar A, Keshavjee S, Louie M, Loeb MB, Brunton J, McGeer AJ, Canadian SARS Research Network, Kelvin DJ (2007) Interferon-mediated immunopathological events are associated with atypical innate and adaptive immune responses in patients with severe acute respiratory syndrome. J Virol 81(16):8692–8706

Channappanavar R, Zhao J, Perlman S (2014) T cell-mediated immune response to respiratory coronaviruses. Immunol Res 59(1–3):118–128. https://doi.org/10.1007/s12026-014-8534-z

Channappanavar R, Fehr AR, Vijay R, Mack M, Zhao J, Meyerholz DK, Perlman S (2016) Dysregulated type I interferon and inflammatory monocyte-macrophage responses cause lethal pneumonia in SARS-CoV-infected mice. Cell Host Microbe 19(2):181–193

Chen YZ, Liu G, Senju S, Wang Q, Irie A, Haruta M, Matsui M, Yasui F, Kohara M, Nishimura Y (2010) Identification of SARS-COV spike protein-derived and HLA-A2-restricted human CTL epitopes by using a new muramyl dipeptide derivative adjuvant. Int J Immunopathol Pharmacol 23(1):165–177

Coleman CM, Sisk JM, Halasz G, Zhong J, Beck SE, Matthews KL, Venkataraman T, Rajagopalan S, Kyratsous CA, Frieman MB (2016) CD8+ T cells and macrophages regulate pathogenesis in a mouse model of Middle East respiratory syndrome. J Virol. 91(1). pii: e01825-16. https://doi.org/10.1128/JVI.01825-16

Drosten C, Günther S, Preiser W et al (2003) Identification of a novel coronavirus in patients with severe acute respiratory syndrome. N Engl J Med 348(20):1967–1976. https://doi.org/10.1056/NEJMoa030747

Fast E, Chen B (2020) Potential T-cell and B-cell epitopes of 2019-nCoV. bioRxiv 2020.02.19.955484. https://doi.org/10.1101/2020.02.19.955484

Fehr AR, Perlman S (2015) Coronaviruses: an overview of their replication and pathogenesis. Methods Mol Biol 1282:1–23. https://doi.org/10.1007/978-1-4939-2438-7_1

Frieman M, Yount B, Heise M, Kopecky-Bromberg SA, Palese P, Baric RS (2007) Severe acute respiratory syndrome coronavirus ORF6 antagonizes STAT1 function by sequestering nuclear import factors on the rough endoplasmic reticulum/Golgi membrane. J Virol 81(18):9812–9824

Hajeer AH, Balkhy H, Johani S, Yousef MZ, Arabi Y (2016) Association of human leukocyte antigen class II alleles with severe Middle East respiratory syndrome-coronavirus infection. Ann Thorac Med 11(3):211–213. https://doi.org/10.4103/1817-1737.185756

Khan S, Siddique R, Shereen MA, Ali A, Liu J, Bai Q, Bashir N, Xue M (2020) The emergence of a novel coronavirus (SARS-CoV-2), their biology and therapeutic options. J Clin Microbiol. pii: JCM.00187-20. https://doi.org/10.1128/JCM.00187-20

Kikkert M (2020) Innate immune evasion by human respiratory RNA viruses. J Innate Immun 12 (1):4–20. https://doi.org/10.1159/000503030

Kopecky-Bromberg SA, Martínez-Sobrido L, Frieman M, Baric RA, Palese P (2007) Severe acute respiratory syndrome coronavirus open reading frame (ORF) 3b, ORF 6, and nucleocapsid proteins function as interferon antagonists. J Virol 81(2):548–557

Kumar S, Maurya VK, Prasad AK et al (2020) Structural, glycosylation and antigenic variation between 2019 novel coronavirus (2019-nCoV) and SARS coronavirus (SARS-CoV). VirusDis 31:13–21. https://doi.org/10.1007/s13337-020-00571-5

Kuri T, Weber F (2010) Interferon interplay helps tissue cells to cope with SARS-coronavirus infection. Virulence 1(4):273–275. https://doi.org/10.4161/viru.1.4.11465

Lau YL, Peiris JS, Law HK (2012) Role of dendritic cells in SARS coronavirus infection. Hong Kong Med J 18(Suppl 3):28–30

Law HK, Cheung CY, Sia SF, Chan YO, Peiris JS, Lau YL (2009) Toll-like receptors, chemokine receptors and death receptor ligands responses in SARS coronavirus infected human monocyte derived dendritic cells. BMC Immunol 10:35. https://doi.org/10.1186/1471-2172-10-35

Lee S, Stokes KL, Currier MG, Sakamoto K, Lukacs NW, Celis E, Moore ML (2012) Vaccine-elicited CD8+ T cells protect against respiratory syncytial virus strain A2-line19F-induced pathogenesis in BALB/c mice. J Virol 86(23):13016–13024. https://doi.org/10.1128/JVI.01770-12

Li F (2016) Structure, function, and evolution of coronavirus spike proteins. Annu Rev Virol 3 (1):237–261

Li G, Chen X, Xu A (2003) Profile of specific antibodies to the SARS-associated coronavirus. N Engl J Med 349(5):508–509

Li CK, Wu H, Yan H, Ma S, Wang L, Zhang M, Tang X, Temperton NJ, Weiss RA, Brenchley JM, Douek DC, Mongkolsapaya J, Tran BH, Lin CL, Screaton GR, Hou JL, McMichael AJ, Xu XN (2008) T cell responses to whole SARS coronavirus in humans. J Immunol 181(8):5490–5500

Li X, Zai J, Zhao Q et al (2020) Evolutionary history, potential intermediate animal host, and cross-species analyses of SARS-CoV-2. J Med Virol 2020:1–10. https://doi.org/10.1002/jmv.25731. [Published online ahead of print, 2020 Feb 27]

Mehta P, McAuley DF, Brown M, Sanchez E, Tattersall RS, Manson JJ; HLH Across Speciality Collaboration, UK (2020) COVID-19: consider cytokine storm syndromes and immunosuppression. Lancet 395(10229):1033–1034. https://doi.org/10.1016/S0140-6736(20)30628-0

Prompetchara E, Ketloy C, Palaga T (2020) Immune responses in COVID-19 and potential vaccines: lessons learned from SARS and MERS epidemic. Asian Pac J Allergy Immunol 38(1):1–9. https://doi.org/10.12932/AP-200220-0772

Sarkar B, Ullah MA, Johora FT, Taniya MA, Araf Y (2020) The essential facts of Wuhan novel coronavirus outbreak in China and epitope-based vaccine designing against COVID-19. bioRxiv 2020.02.05.935072. https://doi.org/10.1101/2020.02.05.935072

Totura AL, Whitmore A, Agnihothram S, Schäfer A, Katze MG, Heise MT, Baric RS (2015) Toll-like receptor 3 signaling via TRIF contributes to a protective innate immune response to severe acute respiratory syndrome coronavirus infection. mBio 6(3):e00638–e00615. https://doi.org/10.1128/mBio.00638-15

Walls AC, Park YJ, Tortorici MA, Wall A, McGuire AT, Veesler D (2020) Structure, Function, and antigenicity of the SARS-CoV-2 spike glycoprotein. Cell 180:1–12. https://doi.org/10.1016/j.cell.2020.02.058. [Epub ahead of print]

Wang SF, Chen KH, Chen M, Li WY, Chen YJ, Tsao CH, Yen MY, Huang JC, Chen YM (2011) Human-leukocyte antigen class I Cw 1502 and class II DR 0301 genotypes are associated with resistance to severe acute respiratory syndrome (SARS) infection. Viral Immunol 24 (5):421–426. https://doi.org/10.1089/vim.2011.0024

Wang Y, Wang Y, Chen Y, Qin Q (2020) Unique epidemiological and clinical features of the emerging 2019 novel coronavirus pneumonia (COVID-19) implicate special control measures. J Med Virol 2020:1–9. https://doi.org/10.1002/jmv.25748. [Epub ahead of print]

Wieczorek M, Abualrous ET, Sticht J, Álvaro-Benito M, Stolzenberg S, Noé F, Freund C (2017) Major histocompatibility complex (MHC) class I and MHC class II proteins: conformational plasticity in antigen presentation. Front Immunol 8:292. https://doi.org/10.3389/fimmu.2017.00292

World Health Organization (2020) Coronavirus disease 2019 (COVID-19) situation report—51. World Health Organization. https://www.who.int/docs/default-source/coronaviruse/situation-reports/20200311-sitrep-51-covid-19.pdf?sfvrsn=1ba62e57_10. Accessed 16 Mar 2020

Yan R, Zhang Y, Li Y, Xia L, Guo Y, Zhou Q (2020) Structural basis for the recognition of the SARS-CoV-2 by full-length human ACE2. Science. pii: eabb2762. https://doi.org/10.1126/science.abb2762

Yao X, Ye F, Zhang M, Cui C, Huang B, Niu P, Liu X, Zhao L, Dong E, Song C, Zhan S, Lu R, Li H, Tan W, Liu D (2020) In vitro antiviral activity and projection of optimized dosing design of hydroxychloroquine for the treatment of severe acute respiratory syndrome coronavirus 2 (SARS-CoV-2). Clin Infect Dis. pii: ciaa237. https://doi.org/10.1093/cid/ciaa237

Yip MS, Leung NH, Cheung CY, Li PH, Lee HH, Daëron M, Peiris JS, Bruzzone R, Jaume M (2014) Antibody-dependent infection of human macrophages by severe acute respiratory syndrome coronavirus. Virol J 11:82. https://doi.org/10.1186/1743-422X-11-82

Yong CY, Ong HK, Yeap SK, Ho KL, Tan WS (2019) Recent advances in the vaccine development against Middle East respiratory syndrome-coronavirus. Front Microbiol 10:1781. https://doi.org/10.3389/fmicb.2019.01781

Yu H, Jiang LF, Fang DY, Yan HJ, Zhou JJ, Zhou JM, Liang Y, Gao Y, Zhao W, Long BG (2007) Selection of SARS-coronavirus-specific B cell epitopes by phage peptide library screening and evaluation of the immunological effect of epitope-based peptides on mice. Virology 359 (2):264–274

Zheng H, Zhang M, Yang C et al (2020) Elevated exhaustion levels and reduced functional diversity of T cells in peripheral blood may predict severe progression in COVID-19 patients. Cell Mol Immunol, 1–3. https://doi.org/10.1038/s41423-020-0401-3

Chapter 6
Clinical Characteristics and Differential Clinical Diagnosis of Novel Coronavirus Disease 2019 (COVID-19)

Raman Sharma, Madhulata Agarwal, Mayank Gupta, Somyata Somendra, and Shailendra K. Saxena

Abstract Novel Coronavirus Disease (COVID-19) has become a rapidly growing pandemic involving several nations. It is of serious concern and extreme challenge not only to the health personnel but also to the countries for containment. The causative organism is SARS-CoV-2, RNA virus of subgenus *Sarbecovirus*, similar to the SARS virus, and seventh member of the human coronavirus family responsible for this zoonotic infection. It binds to the human angiotensin converting enzyme (hACE-2) receptor and causes constitutional and respiratory symptoms. The major mode of transmission is human to human and the median incubation period is 4 days. The most common symptom as studied from various cohorts of COVID-19 patients are fever (83–98%) followed by fatigue (70%) and dry cough (59%); gastrointestinal symptoms are relatively uncommon differentiating it from SARS and MERS. Most of the SAR-CoV-2 infection are mild (80%) with a usual recovery period of 2 weeks. COVID-19 commonly affects males in the middle age and elderly age group, with highest case fatality (8–15%) among those aged >80 years. The disease begins with fever, dry cough, fatigue and myalgia progressing to dyspnoea and ARDS over 6 and 8 days post exposure, respectively. Underlying co-morbidities increase mortality in COVID-19. Poor prognostic factors are elderly, co-morbidities, severe lymphopaenia, high CRP and D-dimer >1 μg/L. The overall mortality rate ranges from 1.5 to 3.6%. COVID-19 has to be differen-

Raman Sharma, Madhulata Agarwal, and Shailendra K. Saxena contributed equally as first author.

R. Sharma (✉) · M. Agarwal · M. Gupta · S. Somendra
Department of Medicine, Sawai Man Singh Medical College, Jaipur, India

S. K. Saxena
Centre for Advanced Research (CFAR)-Stem Cell/Cell Culture Unit, Faculty of Medicine, King George's Medical University (KGMU), Lucknow, India
e-mail: shailen@kgmcindia.edu

S. K. Saxena (ed.), *Coronavirus Disease 2019 (COVID-19)*, Medical Virology: from Pathogenesis to Disease Control, https://doi.org/10.1007/978-981-15-4814-7_6

tiated from other viral and bacterial pneumonias as they are more common among healthy adults. Despite constant and vigorous efforts by researchers and health agencies, we are far from containment, cure or prevention by vaccine; hence right information and stringent prevention and control measures are the only weapon in the armoury to combat the ongoing infection.

Keywords COVID-19 · Clinical features · Differential diagnosis

6.1 Introduction

The world is yet again facing a global pandemic of zoonotic origin after the 2002 SARS (severe acute respiratory syndrome) and 2012 MERS (Middle East respiratory syndrome) era. Renowned microbiologists Macfarlane Burnet and David White predicted in 1972 that "the most likely forecast about the future of infectious diseases is that it will be very dull." They further admitted that there was always a risk of "some wholly unexpected emergence of a new and dangerous infectious disease" (Burnet and White 1972). True to their prediction, in December 2019, a cluster of cases of pneumonia of unknown cause, linked to the local seafood market in Wuhan, capital of Hubei province in China heralded the now global pandemic of SARS-CoV-2, previous called the novel coronavirus 2019 (2019-nCoV) (Wuhan Municipal Health Commission 2019). It was in February 2020 the Director-General of WHO, Dr. Tedros Ghebreyesus, christened it as coronavirus disease 2019 (COVID-19) (World Health Organization 2020a). It was isolated from the human airway epithelial cells, and after the viral genomic sequencing was identified as the seventh member of the *betacoronavirus* family infecting humans, subfamily *Orthocoronavirinae*, order *nidovirales* and subgenus *Sarbecovirus* similar to the SARS virus but different clade with profound similarity to bat coronaviruses (Tan et al. 2020). It is a non-segmented, positive sense, single-stranded RNA virus. After the initial identification of the local cluster of cases in Wuhan, the epidemic peaked in China between late January and early February (WHO 2020a). The virus simultaneously spread rapidly worldwide by community transmission in Italy, Iran, Japan and South Korea and superspreading events affecting many nations across the globe causing the WHO to declare a public health emergency of international concern (PHEIC) on January 30th 2020 and a global pandemic on 12th March 2020 (WHO 2020b). Understanding of the disease is ever evolving since its emergence. SARS-CoV-2 is a pathogen with a naughty mix of virulence and contagiousness combined with human–animal interaction, urban crowding, global trade and tourism has made it a potential threat to mankind worldwide. Rapid containment of this pandemic needs enhancement of our knowledge regarding its transmission, incubation period, reproducibility, pathogenicity and clinical signs and symptoms and the differential diagnosis.

6.2 General Symptoms of Novel Coronavirus Disease 2019 (COVID-19)

COVID-19 has been studied extensively since its first appearance in Wuhan.

6.2.1 Transmission

Majority of the early cases (55%) clustered in Wuhan were initially transmitted due to direct contact with local seafood wholesale market and from residents in its local vicinity, but very soon person-to-person transmission was identified as the major mode of transmission (World Health Organization 2020b). SARS-CoV-2 like SARS virus binds to the human angiotensin-converting enzyme 2 receptor (hACE-2) located on type II alveolar and intestinal cells (Zhou et al. 2020a). Thus, the major mode of transmission is via airborne large droplets with a risk limited to ~6 ft from the patient (del Rio and Malani 2020). Due to large droplets borne transmission, infection can be prevented from spreading by using surgical face masks and airborne precautions. The other mode of person-to-person transmission is contact transmission via fomites, as the large droplets settle on surface and contaminate it, thus mandate decontamination of fomites and good hand hygiene.

6.2.2 Reproduction Number and Secondary Attack Rates

Reproduction number—the number of new cases arising from an infected case—was identified initially at 2·35 and declined to 1·05 over the period from December 2019 to January 2020 after travel restrictions were implemented in Wuhan (Kucharski et al. 2020). Reproduction number >1 implies that the epidemic will increase. Secondary attack rate (SAR), defined as the probability that an infection occurs among susceptible people within a specific group (such as household or close contacts), was identified to be 1–5% in China and 0.45% in the United States (WHO 2020c). These are important in deciding the importance of social interaction in causation of epidemic spread and highlight the significance of contact tracing.

6.2.3 Pathophysiology Behind Symptoms and Signs

An infected person moves through stages of replication over initial few days followed by a stage of adaptive immunity over the next few days (Karakike and Giamarellos-Bourboulis 2019). In the replicative stage the virus replicates, leading to influenza like illness characterized by mild symptoms due to direct cytopathic effect of the virus. In the stage of adaptive immunity viral levels decline as immune system takes over, but the inflammatory cytokine storm leads to tissue destruction and clinical deterioration—which explains the phenomenon where patients remain relatively well initially before deteriorating suddenly. This implicates early initiation of antiviral therapies for better outcomes and use of immunosuppressive therapies in the adaptive immune stage.

6.2.4 Incubation Period (IP)

Incubation period for COVID-19 is defined as the interval between the potential earliest date of contact of the transmission source (wildlife or person with suspected or confirmed case) and the potential earliest date of symptom onset (i.e. cough, fever, fatigue or myalgia) and is within 14 days following exposure. Median incubation period being 4 days (interquartile range, 2–7 days; means 50% cases are dispersed during this period) (Guan et al. 2020).

6.2.5 Demographic Features and General Symptoms of COVID-19

Spectrum of illness ranges from mild disease (no or mild pneumonia) in 81% with a usual recovery period of about 2 weeks, through severe disease (dyspnoea, hypoxia or >50% lung involvement on imaging within 24–48 h) in 14% with a recovery period of about 3–6 weeks to critical disease (ARDS, sepsis, septic shock or MODS) in 5%; this was observed in data of 44,500 confirmed cases of COVID-19 as per information from Chinese Centre for Disease Control and Prevention (Wu and McGoogan 2020). COVID-19 predominantly affects males (58.1%), in particular of Asian ethnicity; this might be due to increased expression of hACE-2 receptor in them and also due to higher rates of smoking which further leads to increased expression of hACE-2 receptor (Cai 2020). But these are just speculations and further studies worldwide involving other ethnicities are needed. The age group affected predominantly are the middle aged and elderly (Median: 47 years; Range: 30–79 years), with very few cases reported among children (0.9–2%). Elderly are more severely affected and have higher case fatality (8–15%) (Zu et al. 2020).

Most studies (summarized in Table 6.1) on hospitalized patients from Wuhan reveal that the common symptoms of COVID-19 are fever (83–98%), fatigue (70%), dry cough (59%), anorexia (40%), myalgia (35%), dyspnoea (31%) and sputum production (27%). Fever in COVID-19 in various cohorts has been described as very low grade (axillary temperature >37.5 °C), intermittent and of prolonged duration (up to ~14 days). In the cohort of 1099 patients, in a study by Guan et al. (2020) only 43.8% had fever on admission and 88.7% had it during hospitalization.

Table 6.1 General symptoms at the time of presentation in various cohorts

	Guan et al. $N = 1099$	Shi et al. $N = 21$	Xu et al. $N = 62$	Zhou et al. $N = 191$	Yang et al. $N = 52$
Constitutional symptoms					
Fever	975 (88.7%)	18 (86%)	48 (77%)	180 (94%)	46 (88%)
Myalgia	164 (14.9%)			29 (15%)	6 (12%)
Headache	150 (13.6%)	2 (10%)	21 (34%)		3 (6%)
Upper respiratory symptoms					
Rhinorrhoea	53 (4.8%)	5 (24%)			3 (6%)
Sore throat	153 (13.9%)				
Lower respiratory symptoms					
Dyspnoea	205 (18.7%)	9 (43%)	2 (3%)	56 (29%)	33 (64%)
Chest tightness		5 (24%)			
Cough	745 (67.8%)	15 (71%)	50 (81%)	151 (79%)	40 (77%)
Sputum	370 (33.7%)	3 (14%)	35 (56%)	44 (23%)	
Haemoptysis	10 (0.9%)		2 (3%)		
Gastrointestinal symptoms					
Nausea/vomiting	55 (5%)	2 (10%)		7 (4%)	2 (6%)
Diarrhoea	42 (3.8%)	1 (5%)	3 (8%)	9 (5%)	

Gastrointestinal symptoms such as nausea and vomiting (5%) and diarrhoea (3.8%) were relatively uncommon. Most common symptoms in most cohorts of COVID-19 are fever and cough, and gastrointestinal symptoms are relatively uncommon which differentiates COVID-19 from SARS and MERS. Asymptomatic infection too was observed but the frequency is unknown as was seen in the *Diamond Princess* cruise ship where out of the 619 people (17%) who were positive for SARS-CoV-2 (Japanese National Institute of Infectious Diseases 2020), 50% were asymptomatic with few of them showing objective abnormalities on imaging. Physical examination is nonspecific and unremarkable with <2% having pharyngitis or tonsillar enlargement.

6.2.6 *COVID-19 in Children and Pregnancy*

Symptomatic infection in children appears to be uncommon and if it does occur it is mild; only rarely severe infection has been documented. In a recent review by Chinese Centre for Disease Control and Prevention, less than 1% of the cases were reported in children less than 10 years of age (Wu and McGoogan 2020). In a study of 1391 children assessed and tested at Wuhan children's hospital from 28th January to 26th February, 2020, a total of 171 (12.3%) were confirmed to have SARS-CoV-2 infection, with 104 males and 67 females. The median age of the children was 6.7 years. Cough was the most common symptom (48.5%), followed by pharyngeal erythema (46.2%) and fever (41.5%). Most common radiological finding was bilateral ground-glass opacities observed in 32.7% children. Radiological features were seen in

12 children with pneumonia without any symptoms and 27 (15.8%) had no radiological findings. As of March 8, 2020, 29 are stable in ward and 149 have been discharged with only single mortality of a child with underlying intussusceptions (Lu et al. 2020). In a large cohort study from China, <2% infection by SARS-COV-2 occurs in young individuals <20 years (Wu and McGoogan 2020).

There is very little data regarding COVID-19 in pregnancy, and knowledge regarding its intra-uterine and perinatal transmission is scarce. However, in a study of 18 pregnant females including both confirmed and suspected cases of SARS-CoV-2 infection only two cases of neonatal transmission were reported. One was by contact infection from the mother (Chen et al. 2020a). Thus, as of now the diagnosis and management of pregnant individuals is same as non-pregnant females. It is unknown if transmission occurs via breast milk; hence, airborne droplet and contact precautions are mandatory in lactating mothers or they have to avoid breastfeeding. So far the outcome studies in pregnancy seem inadequate due to the prolonged period of gestation.

6.3 Clinical Course of COVID-19 (Day-Wise)

The symptoms of COVID-19 (Fig. 6.1) initially begin with symptoms of fatigue, low-grade intermittent fever of prolonged duration, myalgia, dry cough and shortness of breath, which then either improves with early identification and conservative management (Fig. 6.2) or worsens and progresses to dyspnoea and productive cough. The median time to onset of dyspnoea from various cohorts was found to

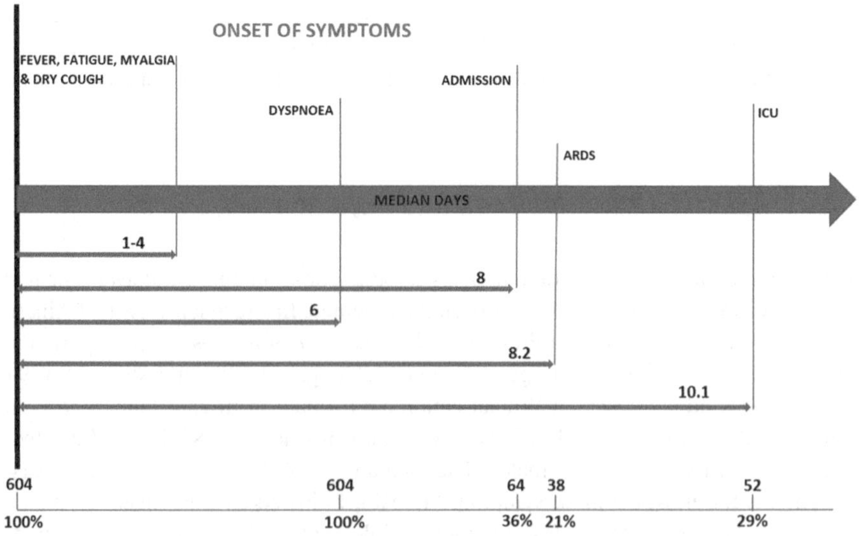

Fig. 6.1 Timeline of COVID-19 after onset of illness

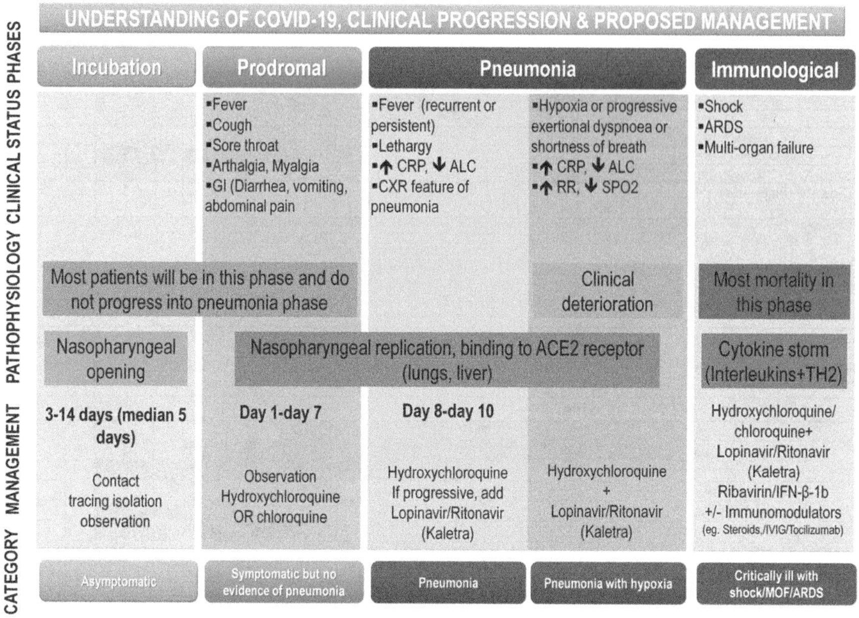

Fig. 6.2 Clinical progression and management of COVID-19

Table 6.2 Summary of course of illness in days in various cohorts of COVID-19

Variables	Zhou et al. $N = 199$	Huang et al. $N = 41$
Onset following exposure in days		
Dyspnoea	13	8
Admission		7
ARDS	12	9
Mechanical ventilation/ICU	14.5	10.5

be 6 days following exposure. The median time to admission, development of ARDS and need for mechanical ventilation and ICU care was 8, 8.2 and 10 days, respectively. In a study (summarized in Table 6.2) by Zhou et al. (2020b) the median time to discharge from onset of illness was 22 days (IQR: 18–25 days), whereas the median time to death was 18·5 days (15–22 days). The median duration of fever was 12 days (8–13 days) and cough persisted for 19 days (IQR: 12–23 days) in survivors. Complications such as bilateral pneumonia followed by ARDS, sepsis and septic shock, acute cardiac injury, acute kidney injury and secondary infection developed at a median of 12 days (8–15 days), 9 days (7–13 days), 15 days (10–17 days), 15 days (13–19 days) and 17 days (13–19 days). The median duration of viral shedding was 20 days (IQR 17–24 days) from illness onset in survivors (Fig. 6.3a) vs. non-survivors (Fig. 6.3b) who continued shedding virus till death. In patients who received anti-retroviral with a median time of 14 days to their

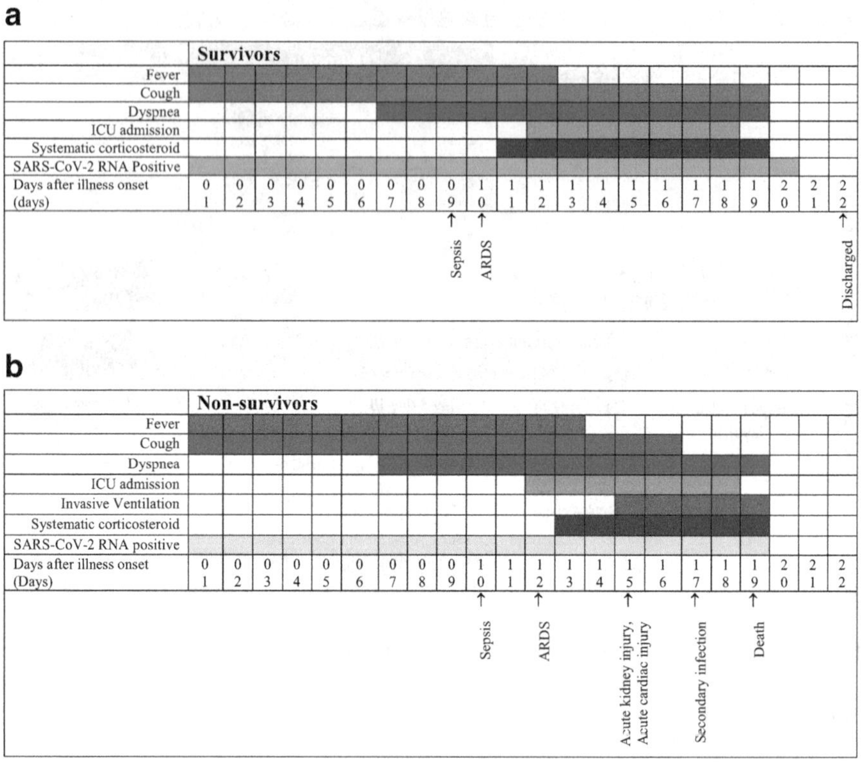

Fig. 6.3 (**a**) Course of illness among COVID-19 survivors. (**b**) Course of illness among COVID-19 non-survivors

initiation, the median duration of viral shedding was 22 days (18–24 days). The median duration of viral shedding was 19 days (17–22 days) and 24 days (22–30 days) in patients with severe disease and critical disease, respectively.

6.4 Clinical Complications and Case Definitions

6.4.1 Clinical Complications and Outcomes

The most common complications that develop in COVID-19 are bilateral pneumonia which may progress to ARDS, sepsis and septic shock, acute kidney injury (AKI) and others such as acute cardiac injury (arrhythmias, heart failure, MI), coagulopathy, rhabdomyolysis, hyponatremia and acidosis. Complications are more in severe disease vs. non-severe disease. In the study cohort of 1099 COVID-19 cases by Guan et al. (2020), bilateral pneumonia (91.1%) occurred

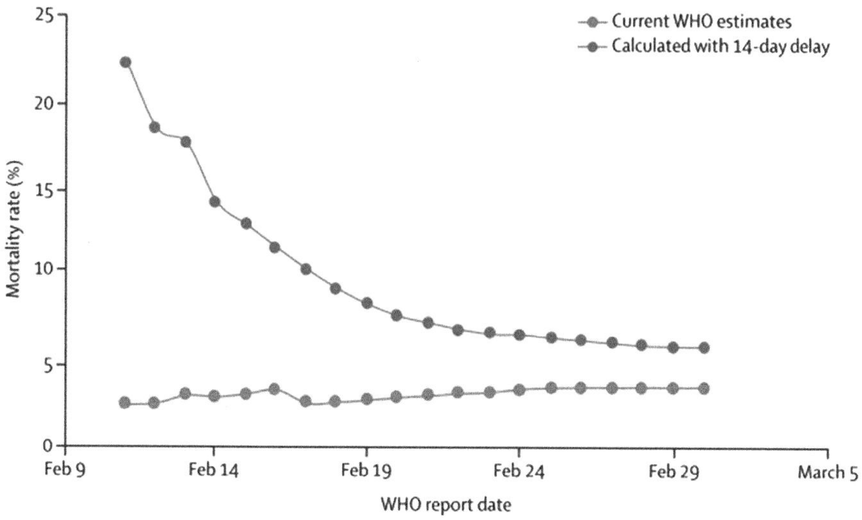

Fig. 6.4 Current global mortality rates estimate of COVID-19 by the WHO (February 11 to March 1, 2020)

most commonly followed by ARDS (3.4%), septic shock (1.1%), AKI (0.5%) and coagulopathy (0.1%). In this cohort 926 (84%) had non-severe and 173 (16%) had severe disease. Underlying co-morbidities were more common among patients with severe disease than among non-severe disease (38.7% vs. 21.0%). Outcomes were as follows: 55 (5%) were discharged, 15 (1.4%) died, 9 (0.8%) recovered and 1029 (93.6%) remain hospitalized as of March 6, 2020. Overall mortality associated with COVID as of March 1, 2020, as estimated by the WHO, is 3.6% in China and 1.5% outside China. The mortality of COVID-19 (Fig. 6.4) is much less than that of SARS (9.6%) and MERS (34%). Case fatality rate of COVID-19 ranges from 5.8% in Wuhan to 0.7% in China. It is higher in elderly >70 years (8–15%). Overall case fatality is 2.3% (WHO 2020c). In most of the people with COVID-19, almost >80% have mild illness (Wu and McGoogan 2020).

6.4.2 *Underlying Co-morbidities Complicating COVID-19*

The major underlying co-morbidities that complicate the course of COVID-19 by increasing the severity of illness, use of mechanical ventilation and length of ICU stay and thus increase the mortality include uncontrolled hypertension, diabetes, coronary heart disease, hepatitis B, cerebrovascular disease, chronic obstructive airway disease and others like cancer, chronic kidney disease and immunodeficiency. In a study by Guan et al. (2020), hypertension (15%) was predominantly followed by diabetes (7.4%), the probable explanation to this could be due to hACE-2 receptor polymorphism in Asian population (Table 6.3).

Table. 6.3 Summarized co-morbidities, complications and mortality in various cohorts of COVID-19

Variables	Guan et al. $N = 1099$ (%)	Zhou et al. $N = 191$ (%)	Huang et al. $N = 41$ (%)
Hypertension	15	30	15
Diabetes	7.4	19	20
Coronary heart disease	2.5	8	15
Chronic obstructive pulmonary disease	1.1	3	2
Cerebrovascular disease	1.4		
Hepatitis B	2.1		
Cancer	0.9	1	2
Pneumonia	91.1		
Septic shock	1.1	20	7
ARDS	3.4	31	29
Acute kidney injury	0.5	15	7
Acute cardiac injury		17	12
Coagulopathy	0.1	19	
Rhabdomyolysis	0.2		
Invasive mechanical ventilation	2.3	17	5
ECMO	5	2	5
CRRT	9	5	7
Mortality	1.4	28.3	15

6.4.3 *Case Definitions (WHO 2020d)*

Surveillance definition: A person with acute respiratory infection (sudden onset of at least one of the following: fever, cough, sore throat, shortness of breath) requiring hospitalization or not

AND

In the 14 days prior to symptom onset, meets at least one of the following epidemiological criteria:

- Was in close contact with a confirmed or probable case of COVID-19
- History of travel to areas of China with ongoing community transmission of SARS-CoV-2
- Worked in or attended a health care facility where COVID-19 patients were being treated

Probable case: A suspected case in whom testing for SARS-CoV-2 is inconclusive (result of the test reported by BSL4 lab) or in whom testing was positive on a pan-coronavirus assay.

Confirmed case: A person with laboratory confirmation of virus causing COVID-19 infection by RT-PCR of oropharyngeal or nasopharyngeal swab, irrespective of clinical signs and symptoms.

6.5 Clinical Diagnosis of Novel Coronavirus Disease 2019 (COVID-19)

Clinical diagnosis is based on signs and symptoms as described in previous sections along with substantiation from routine laboratory investigations and imaging and further confirmed by RT-PCR of nasopharyngeal or oropharyngeal swabs.

6.5.1 Routine Pathology and Biochemistry Lab Findings and Imaging in COVID-19

Major lab findings (Table 6.4) as in all other respiratory viral illness include leukopenia and lymphopaenia, elevated transaminases and D-dimer (Centres for Disease Control and Prevention 2020). Severe lymphopaenia, elevated C-reactive protein (CRP), elevated D-dimer (>1 μg/L), IL-6, ALT, serum ferritin, lactate dehydrogenase, creatine kinase, high-sensitivity cardiac troponin I, creatinine, prothrombin time and procalcitonin are associated with higher mortality (Guan et al. 2020).

The most common radiological findings on X-ray and CT thorax observed are bilateral ground-glass opacities (GGOs) with or without consolidation. Lesions predominantly involve bilateral lower lobes with peripheral distribution (Shi et al. 2020).

Table. 6.4 Laboratory findings in patients of COVID-19 in various cohorts

Variables	Guan et al. *N* = 1099 (%)	Zhou et al. *N* = 191 (%)	Huang et al. *N* = 41 (%)
Lymphopaenia <0.8 × 10^9/L	83.2	40	63
Thrombocytopenia <100 × 10^9/L	36.2	7	5
Elevated ALT >40 U/L	21.3	31	37
Elevated D-dimer >1 μg/L	46.4	42	
Elevated CRP	60.7		
Elevated creatinine	1.6	4	10
Elevated LDH >245 U/L	41	67	73
Elevated IL-6 >7.4 pg/mL			
Serum ferritin >300 μg/L		80	
Elevated creatinine kinase >185 U/L	13.7	13	33
High-sensitivity troponin I >28 pg/mL		17	12
Increased procalcitonin ≥0.5 ng/mL	5.5	9	8
Increased prothrombin time ≥16 s		6	

6.5.2 *Poor Prognostic Factors in COVID-19*

Strong independent predictors of high mortality are elderly (age ≥70 years); underlying co-morbidities such as uncontrolled hypertension, diabetes and coronary artery disease, chronic obstructive pulmonary disease and malignancy; severe lymphopaenia ($<0.8 \times 10^9$/L) and D-dimer (>1 μg/L) (Chen et al. 2020b). Other poor prognostic factors are elevated C-reactive protein, LDH, ALT, serum ferritin, IL-6 and high-sensitivity cardiac troponin.

6.6 Differential Diagnosis of Novel Coronavirus Disease 2019 (COVID-19)

COVID-19 needs to be differentiated from other respiratory illnesses (Fig. 6.5) as well as other viral pneumonias (Table 6.5) associated with adenovirus, influenza, human metapneumovirus, parainfluenza, respiratory syncytial virus (RSV), rhinovirus and bacterial pneumonias (Ishiguro et al. 2019).

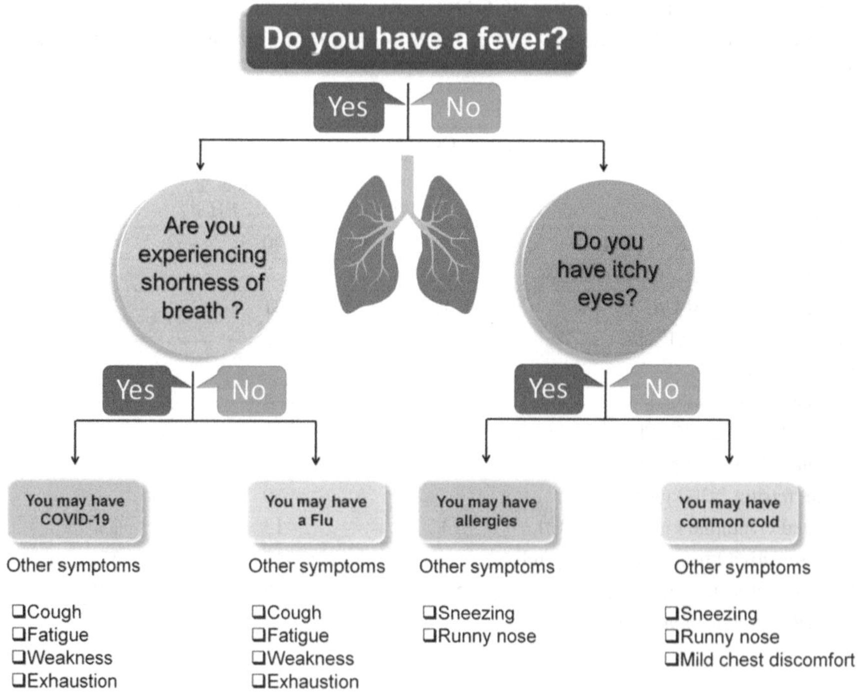

Fig. 6.5 Flow chart depicting general symptoms of COVID-19 and its mimics

Table 6.5 Manifestations of COVID-19 and other viral and bacterial pneumonia

Variables	Viral pneumonia	Bacterial pneumonia	COVID-19
Pathogen	Adenovirus, influenza A & B, human metapneumovirus, parainfluenza, respiratory syncytial virus (RSV), rhinovirus, cytomegalovirus	Streptococci, mycoplasma, chlamydia, legionella	SARS-CoV-2
History of exposure to COVID-19	Common in children in winter and spring, uncommon in adults or community	Common in winter, common in both children and adults	History of exposure to Wuhan or areas of ongoing community transmission like Iran, Italy, Japan, South Korea. Affects middle aged and elderly
First symptoms	High-grade fever, cough, sore throat, myalgia	Nasal obstruction, rhinorrhoea, sore throat; symptoms are usually mild	Low-grade fever and dry cough are predominant symptoms
Laboratory findings	RT-PCR positive for the underlying virus, elevated lymphocyte counts	High leucocyte count, elevated ESR and CRP	RT-PCR positive for SARS-CoV-2, lymphopaenia, elevated aminotransferases, CRP and D-dimer
Chest CT findings	Interstitial inflammation, high-attenuation reticular patterns, localized atelectasis or pulmonary oedema	Bronchial or lobar pneumonia, bronchial wall thickening, multiple consolidation patches and centrilobular nodules	Early stage: GGOs Progressive stage: multiple GGOs, consolidation patches, crazy-pavement pattern Advanced stage: diffuse exudative lesions, white-out lung

6.7 Conclusions and Future Perspectives: What We Do Not Know Today?

Novel coronavirus disease (COVID-19) presents as mild illness in majority of cases with <5% developing life-threatening critical illness. Thus, timely recognition of appropriate history of exposure and prompt recognition of symptoms will help identify cases early and help in better contact tracing and early isolation. This will help bring down superspreading events and prevent further spread of the disease, also it will aid in reducing mortality related to COVID-19.

Since its origin in Wuhan with a cluster of cases of pneumonia of unknown origin, a long path has been tread in research on SARS-CoV-2. Our knowledge regarding this novel human coronavirus grows with each passing day, due to untiring efforts of researchers, epidemiologists and various health agencies, yet we are far from the magic bullet. There are lots of questions that remain unanswered, like its zoonotic origin, the frequency of asymptomatic cases, duration of viral shedding in cases, period of infectivity, pathogenicity, long-term sequelae of the disease, its consequences in pregnant individuals, development of a vaccine, etc. In order to contain the pandemic effectively, the epidemiological questions that need answers are regarding; shape of the disease pyramid, percentage infected who develop the disease and proportion of those who seek health care. Amidst all the uncertainty, the pandemic of SARS-CoV-2 grows unabated, fuelled by misinformation causing panic worldwide. This information has to be counteracted by authentic scientific narration. History has lot of lessons to offer regarding mitigating large epidemics, provided we be wise and combat the epidemic by strict vigilance and surveillance, stringent prevention and control practices and intervention strategies. In the era of globalization and mass media, dissemination of life saving research can alone aid in curbing the monstrously growing global epidemic of COVID-19.

Executive Summary

- SARS-CoV-2 was isolated from the human airway epithelial cells and was identified as the seventh member of the *betacoronavirus* family infecting humans, subfamily *Orthocoronavirinae* and subgenus *Sarbecovirus* similar to the SARS virus but different clade with profound similarity to bat coronaviruses.
- Clinical features of COVID-19
  - Transmitted mainly by human to human transmission and by contact with infected fomites.
  - Has a median incubation period of 4 days (Interquartile range: 2–7 days).
  - It predominantly affects middle aged and elderly males with a high case fatality of 8–15% in age group >70 years, affects <2% individuals <20 years.
  - The common symptoms are fever (low grade) in (83–98%), fatigue (70%) and dry cough (59%); GI symptoms are uncommon.
  - Most of the cases are of mild infection (80%) with a usual recovery period of 2 weeks; only 15 and 5% of cases are of severe and critical nature respectively.
- Clinical course of COVID-19
  - The disease begins with fever and dry cough and progresses to dyspnoea and ARDS and need for mechanical ventilation over 6, 8.2, and 10 days post exposure, respectively.
- Complications, outcome and case definition:
  - Most common complications are bilateral pneumonia progressing to ARDS, sepsis and septic shock.

(continued)

- Major underlying co-morbidities that complicate the course of COVID-19 include hypertension, diabetes, coronary heart disease, chronic obstructive pulmonary disease and malignancy.
- Overall mortality rate associated with COVID-19 ranges from 1.5 to 3.6%. The mortality rate of COVID-19 is much less than that of SARS (9.6%) and MERS (34%). Case fatality of COVID-19 is 2.3%.
- Confirmed case is defined as a person with laboratory confirmation of virus causing COVID-19 infection, irrespective of clinical signs and symptoms.

- Clinical diagnosis of COVID-19 and prognostic factors
  - Clinical diagnosis is based on signs and symptoms along with routine laboratory investigations and imaging and confirmed by RT-PCR of nasopharyngeal or oropharyngeal swabs.
  - Major lab findings as in all other respiratory viral illness include leukopenia and lymphopaenia, elevated transaminases, CRP and D-dimer.
  - Most common radiological findings on X-ray and CT thorax observed are bilateral ground-glass opacities (GGOs), patchy consolidation or extensive exudative infiltrates.
  - Strong independent predictors of high mortality are elderly (age $\geq$80 years); underlying co-morbidities such as uncontrolled hypertension, diabetes and coronary artery disease, chronic obstructive pulmonary disease and malignancy; severe lymphopaenia ($<0.8 \times 10^9$/L) and D-dimer (>1 μg/L).
- Differential diagnosis includes other viral pneumonia like adenovirus, influenza, human metapneumovirus, parainfluenza, respiratory syncytial virus (RSV), rhinovirus and bacterial pneumonia.

References

Burnet M, White DO (1972) Natural history of infectious disease, 4th edn. Cambridge University Press, Cambridge

Cai G (2020) Bulk and single-cell transcriptomics identify tobacco-use disparity in lung gene expression of ACE2, the receptor of 2019-nCov. MedRxiv. Published online Feb 28. https://doi.org/10.1101/2020.02.05.20020107

Centres for Disease Control and Prevention (2020) Interim clinical guidance for management of patients with confirmed 2019 novel coronavirus (2019-nCoV) infection, updated February 1 2, 2020. https://www.cdc.gov/coronavirus/2019-ncov/hcp/clinical-guidance-management-patients.html. Accessed 14 Feb 2020

Chen H, Guo J, Wang C et al (2020a) Clinical characteristics and intrauterine vertical transmission potential of COVID-19 infection in nine pregnant women: a retrospective review of medical records. Lancet 395:809–815

Chen N, Zhou M, Dong X et al (2020b) Epidemiological and clinical characteristics of 99 cases of 2019 novel coronavirus pneumonia in Wuhan, China: a descriptive study. Lancet 395:507

del Rio C, Malani PN (2020) COVID-19—new insights on a rapidly changing epidemic. JAMA. Published online February 28, 2020. https://doi.org/10.1001/jama.2020.3072
Guan WJ, Ni ZY, Hu Y et al (2020) Clinical characteristics of coronavirus disease 2019 in China. N Engl J Med
Kucharski AJ, Russell TW, Diamond C, Liu Y, Edmunds J, Funk S (2020) Early dynamics of transmission and control of COVID-19: a mathematical modelling study. Lancet Infect Dis. https://doi.org/10.1016/S1473-3099(20)30144-4
Ishiguro T, Kobayashi Y, Uozumi R et al (2019) Viral pneumonia requiring differentiation from acute and progressive diffuse interstitial lung diseases. Intern Med 58(24):3509–3519
Japanese National Institute of Infectious Diseases (2020) Field briefing: diamond princess COVID-19 cases, 20 Feb Update. https://www.niid.go.jp/niid/en/2019-ncov-e/9417-covid-dp-fe-02.html. Accessed 1 Mar 2020
Karakike E, Giamarellos-Bourboulis EJ (2019) Macrophage activation-like syndrome: a distinct entity leading to early death in sepsis. Front Immunol 10:55
Lu X, Zhang L, Du H et al (2020) SARS-CoV-2 infection in children. N Engl J Med
Shi H, Han X, Jiang N et al (2020) Radiological findings from 81 patients with COVID-19 pneumonia in Wuhan, China: a descriptive study. Lancet Infect Dis 20:425–434
Tan WJ, Zhao X, Ma XJ et al (2020) A novel coronavirus genome identified in a cluster of pneumonia cases—Wuhan, China 2019−2020. China CDC Weekly 2:61–62
WHO (2020a) Director-General's opening remarks at the media briefing on COVID-19—24 February 2020. https://www.who.int/dg/speeches/detail/who-director-general-s-opening-remarks-at-the-media-briefing-on-covid-19%2D%2D-24-february-2020. Accessed 26 Feb 2020
WHO (2020b) Coronavirus disease 2019 (COVID-19) situation report—36. February 25, 2020. https://www.who.int/docs/default-source/coronaviruse/situation-reports/20200225-sitrep-36-covid-19.pdf?sfvrsn=2791b4e0_2. Accessed 26 Feb 2020
WHO (2020c) Report of the WHO-China Joint Mission on coronavirus disease 2019 (COVID-2019). February 16–24, 2020. http://www.who.int/docs/default-source/coronaviruse/who-china-joint-mission-on-covid-19-final-report.pdf. Accessed 4 Mar 2020
WHO (2020d) Coronavirus disease (COVID-19) technical guidance: surveillance and case definitions. https://www.who.int/emergencies/diseases/novel-coronavirus-2019/technical-guidance/surveillance-and-case-definitions. Accessed 28 Feb 2020
World Health Organization (2020a) Director-General's remarks at the media briefing on 2019-nCoV on 11 February 2020. https://www.who.int/dg/speeches/detail/who-director-general-s-remarks-at-the-media-briefing-on-2019-ncov-on-11-february-2020. Accessed 12 Feb 2020
World Health Organization (2020b) Novel coronavirus situation report—2. January 22, 2020. https://www.who.int/docs/default-source/coronaviruse/situation-reports/20200122-sitrep-2-2019-ncov.pdf. Accessed 23 Jan 2020
Wu Z, McGoogan JM (2020) Characteristics of and important lessons from the coronavirus disease 2019 (COVID-19) outbreak in China: summary of a report of 72 314 cases from the Chinese Centre for Disease Control and Prevention. JAMA
Wuhan Municipal Health Commission (2019) Report of clustering pneumonia of unknown aetiology in Wuhan City. Wuhan Municipal Health Commission. http://wjw.wuhan.gov.cn/front/web/showDetail/2019123108989
Zhou P, Yang X-L, Wang X-G et al (2020a) Discovery of a novel coronavirus associated with the recent pneumonia outbreak in 2 humans and its potential bat origin. BioRxiv
Zhou F, Yu T, Du R et al (2020b) Clinical course and risk factors for mortality of adult inpatients with COVID-19 in Wuhan, China: a retrospective cohort study. Lancet 395:1054–1062
Zu F, Yu T, Du R et al (2020) Clinical course and risk factors for mortality of adult inpatients with COVID-19 in Wuhan, China: a retrospective cohort study. Lancet

Chapter 7
Coronavirus Infection Among Children and Adolescents

Sujita Kumar Kar, Nishant Verma, and Shailendra K. Saxena

Abstract Coronavirus infection is a global emergency. Over the past few months, there is a rapid increase in the number of cases and deaths due to coronavirus infection. It has been observed that elderly individuals and those with medical co-morbidities are maximally affected. In children and adolescents, coronavirus infection has low mortality as well as the severity of symptoms are less. Children and adolescents with immunocompromised state, malnutrition, medical co-morbidities and poor hygiene are at higher risk of contracting coronavirus infection. Minimizing this risk factors and adopting appropriate prevention measures will be helpful in limiting the spread of infection as there is no specific treatment and immunization available to date to address this serious issue. This chapter highlights the issues and challenges of coronavirus infection in children and adolescents.

Keywords Coronavirus infection · Children · Adolescents · Prevention

Sujita Kumar Kar and Shailendra K. Saxena contributed equally as first author.

S. K. Kar (✉)
Department of Psychiatry, Faculty of Medicine, King George's Medical University (KGMU), Lucknow, India

N. Verma
Department of Pediatrics, Faculty of Medicine, King George's Medical University (KGMU), Lucknow, India

S. K. Saxena
Centre for Advanced Research (CFAR)-Stem Cell/Cell Culture Unit, Faculty of Medicine, King George's Medical University (KGMU), Lucknow, India
e-mail: shailen@kgmcindia.edu

S. K. Saxena (ed.), *Coronavirus Disease 2019 (COVID-19)*, Medical Virology: from Pathogenesis to Disease Control, https://doi.org/10.1007/978-981-15-4814-7_7

7.1 Introduction

Novel coronavirus (COVID-19) infection, recently declared as a pandemic by the World Health Organization (WHO), is a global threat. According to the situation report of the WHO, by the end of the third week of March 2020, there are more than 266,000 confirmed cases and 11,184 deaths over 182 countries globally (WHO 2020). The above data depict the grievousness of the endemic in the contemporary world. People of all ages are affected with the deadly coronavirus infection, though the risk of infection is higher among older people and those with medical co-morbidities (Wu et al. 2020a). As per the initial report from Wuhan, China, less number of children were affected with coronavirus infection than adult patients (Liu et al. 2020).

7.2 Coronavirus Infection Among Children

As per the WHO report in February 2020, there was no fatality due to coronavirus infection in the 0–9 years age group and the death rate was 0.2% in the age group of 10–19 years; however in the age group of 70–79 years and >80 years the death rates are 8.0% and 14.8%, respectively (World Health Organization 2020; Worldometers 2020). Though children and adolescents have lower death rate, they can be potential agents of transmission. As the degree of mobility is higher in this age group of population, the probability of contracting infection and transmitting to others (particularly the high-risk population of elderly) is high. Children infected with coronavirus often present with fever, cough and breathing difficulties. Many children also report vomiting, diarrhoea and the aforementioned symptoms (Yang et al. 2020). Children born from infected mothers have higher risk of contracting the infection from mother (Yang et al. 2020). The clinical symptoms and investigation findings in children with coronavirus infection often resemble with viral pneumonia (Yang et al. 2020). It has been reported that children often have milder symptoms than adults and the elderly (Chen et al. 2020a; American Academy of Pediatrics 2020; Lu and Shi 2020). The biochemical changes and chest computed tomography (CT) changes of children with coronavirus infection also differ from those of adults (Xia et al. 2020). Children often have more upper respiratory infection than lower respiratory infection, which increases their ability to transmit the infection (American Academy of Pediatrics 2020). A report mentions that even an apparently healthy infant had heavy viral load which indicates that children can even transmit infection without manifesting the illness (Kam et al. 2020).

7.3 Risk Factors Specific to Children and Adolescents

7.3.1 Immunocompromised State

Immunocompromised state is a potential risk factor for acquiring highly contagious infections like novel coronavirus infection. Children and adolescents with poor immune function or on immune-suppressant medications need to be cautious. Preterm babies and newborn babies with low birth weight are also at risk due to their immunocompromised state.

7.3.2 Malnutrition

Malnutrition is still a common problem in the developing and underdeveloped countries. Children with protein-energy malnutrition (Marasmus and Kwashiorkor) or specific vitamin and micronutrient deficiency are at risk of acquiring infections due to their poor body immunity.

7.3.3 Medical Co-morbidities

Specific medical co-morbidities increase the risk of coronavirus infection. Children and adolescents with cardiac disease (mostly congenital heart diseases) and respiratory diseases (bronchial asthma, bronchiectasis) are more vulnerable as coronavirus mostly infects the respiratory system. Patients with haematological disorders like anaemia, leukaemia, etc. also have a compromised immune system, which makes them vulnerable to coronavirus infection.

7.3.4 Poor Hygiene

Small children are often dependent on their cares for personal hygiene. Lack of understanding about the importance of personal hygiene makes them vulnerable to acquire infections.

7.3.5 Lack of Sensitization

Small children are not aware of the concept of pandemic, its seriousness and importance of all preventive measures recommended. Unfortunately, the sensitization activities run by various government and non-government agencies mostly target adults and youths. Many parents seldom discuss or explain the issue of coronavirus infection with their children. All the above factors result in improper sensitization of children as a result of which they are at risk of contracting as well as spreading infection.

7.3.6 Age-Specific Issues

Children are often playful. They talk loudly and express themselves without restraints. Evidence suggests that talking loudly and shouting may cause the spread of the infection through droplets (Chen et al. 2020a). Similarly, touching the face, nose and mouth with hand is common during play among children. It also increases the risk of transmission of coronavirus infection (Chen et al. 2020a). Similarly, children often spend a significant proportion of their time out of home (in school, play activities) and may come in contact with individuals with coronavirus infection.

7.4 Prevention of Coronavirus Infection Among Children and Adolescents

There is no specific treatment for coronavirus infection to date. The evidences gathered in favour of certain anti-retroviral agents and antimalarial agents are not robust yet. Patients with coronavirus infection need to be treated symptomatically and monitoring needs to be done for organ failures. As there is no definite treatment to date for this illness, prevention becomes the top priority.

The Centre for Disease Control and Prevention had issued certain instructions in the public interest that intends to create awareness among the public about coronavirus infection and its prevention among children (Centers for Disease Control and Prevention 2020). As small children may not be able to take their own responsibility, parents, teachers and sensitive citizens of this civilized society should take responsibility to prevent the spread of coronavirus infection. They need to monitor the activities of children at home, school and outside the home setting. There is a need to restrict large group activities, limit play time and keep distance during play and interaction. Similarly in the home and school setting, there is a need to keep the surfaces and objects (walls, toilets, chairs, tables, boards, play items, reading materials) sanitized through repeated cleaning as these remain the medium of transfer of pathogens from infected individuals to healthy ones (Centers for Disease Control and Prevention 2020). Choosing outdoor game in small groups may be more beneficial than indoor games as the ventilation is better outside and the possibility of maintaining distance during play is higher in outdoor play. Group travel, picnics and study tours are to be strictly discouraged. Children need to be taught about hygiene regularly, and they need to be monitored for the implementation of hygiene in practice.

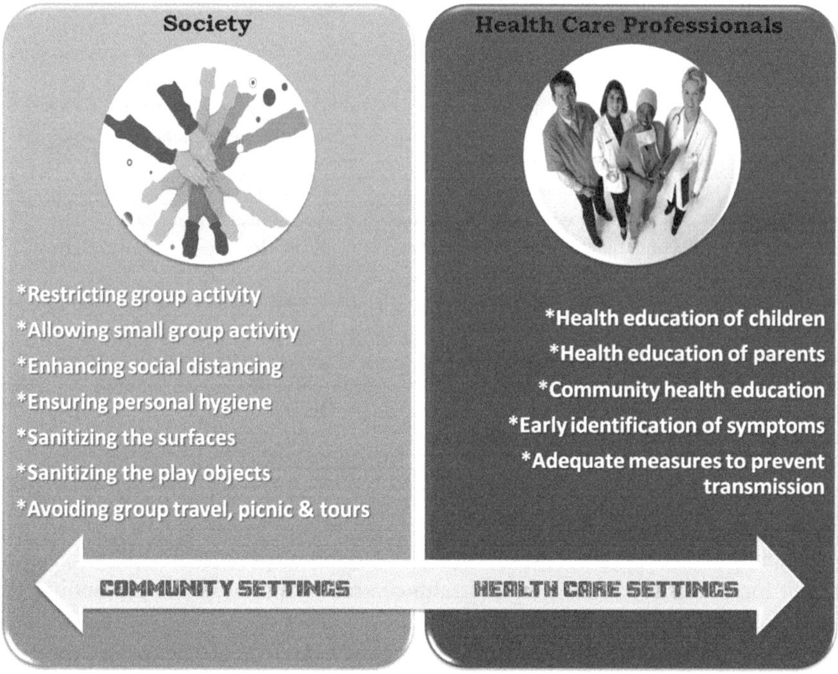

Fig. 7.1 Recommended preventive measures against coronavirus infection in children

Healthcare professionals on the other hand have a pivotal role in providing health education to child and parents and conducting awareness camps in schools and community, as well as regular health check-up for the early identification and prompt treatment of health ailments. Figure 7.1 provides a summary of preventive measures against coronavirus infection in children.

It is of paramount importance to ensure that children should properly sanitize themselves before coming in contact with other family members (particularly elderly and those who have medical illnesses) to limit the possible transmission of coronavirus infection.

Older children and adolescents are trainable and educable about the basics of hygiene and its relevance in the context of coronavirus infection. The above recommendations also stand valid for older children and adolescents for the prevention of coronavirus infection transmission.

Additionally children should avoid contact with persons or other children with recent travel history or contact with persons with recent travel history or those with respiratory infections or fever (Chen et al. 2020a). It is important to target various risk factors (predisposing factors, precipitating factors and perpetuating factors) to limit the chances of getting infected (Kar and Tripathy 2019). Table 7.1 summarizes the potential risk factors that can be targets of intervention.

Table 7.1 Potential risk factors, which can be targets of intervention

Nature of risk factors	Examples	Preventive measures
Predisposing factors	• Low immunity • Malnutrition • Medical co-morbidity • Poor hygiene	• Dietary supplementation • Adequate treatment of medical co-morbidity • Hygiene
Precipitating factors	• Contact with infected persons, contaminated surfaces and objects	• Social distancing • Restricting play, tour, travel, picnic, etc. • Proper sanitation
Perpetuating (maintaining) factors	• Low immunity • Malnutrition • Medical co-morbidity • Poor hygiene	• Dietary supplementation • Adequate treatment of medical co-morbidity • Hygiene • Early identification, isolation and prompt treatment

Children and adolescents, who develop fever and respiratory infections, need to consult for evaluation at the nearest health centres with appropriate precautions till coronavirus infection is ruled out.

7.5 Evidence-Based Management Approach

There is no specific treatment for novel coronavirus disease (COVID-19) supported with evidence to date. Patients with coronavirus infection are often given symptomatic treatment and supportive care (Wu et al. 2020a, b). However, probable management approach against coronavirus infection in children has been exhibited in Fig. 7.2. Researchers found the possible roles of hydroxychloroquine, anti-retroviral medications and interferon in the management of coronavirus infection (Yang et al. 2020). Antibiotics are recommended for secondary bacterial infection and pneumonia. Corticosteroids are to be avoided, except exceptional situations like septic shock, rapidly deteriorating chest imaging, and presence of obvious toxic symptoms like encephalitis or encephalopathy (Chen et al. 2020a).

There is no specific vaccine available for the prevention of COVID-19 infection. Many vaccine trials are going on globally; however, it has been recommended that uninfected people and health workers need to get vaccinated for influenza (Zhang and Liu 2020). There is a possible role of convalescent plasma (if available) in the management of COVID-19 infection (Zhang and Liu 2020).

As of now, prevention is the best option for controlling the rapidly spreading infection of coronavirus. It has been recommended that newborn babies of mothers infected with COVID-19 need to be isolated immediately after delivery to prevent them from acquiring infection (Yang et al. 2020). However, there is no evidence that

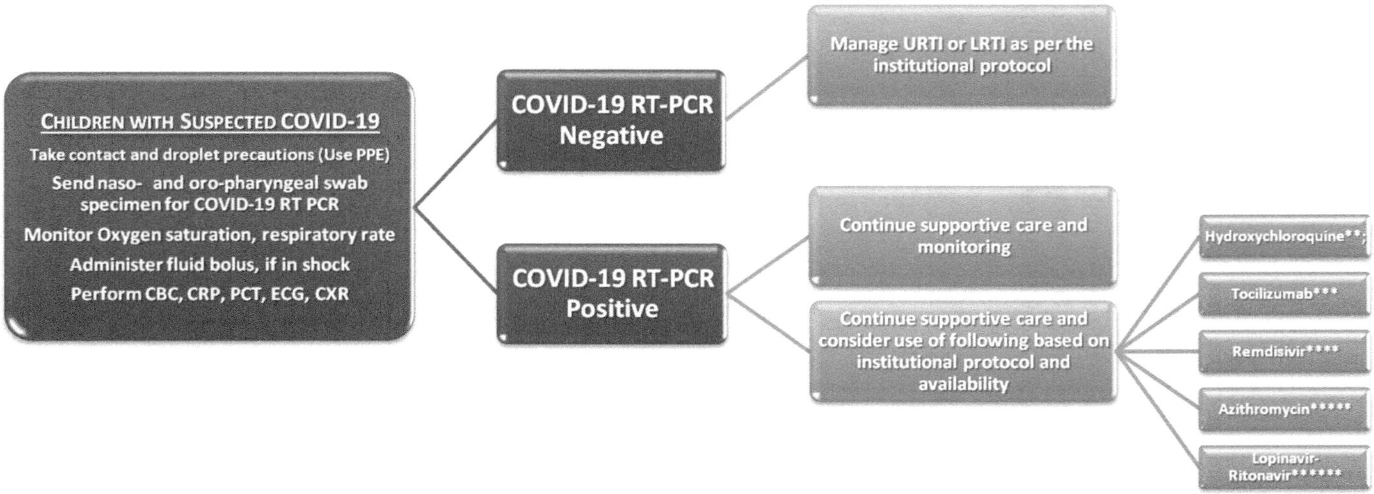

Fig. 7.2 Current probable management approach against coronavirus infection in children. *PPE* personal protective equipment, *CBC* complete blood count, *CRP* C-reactive protein, *PCT* procalcitonin, *ECG* electrocardiogram, *CXR* chest X-ray, *URTI* upper respiratory tract infection, *LRTI* lower respiratory tract infection. *Worsening: Increasing respiratory distress, worsening gas exchange, hypoxia, radiological deterioration. **Hydroxychloroquine (HCQ): Dose 10–13 mg/kg (max: 600 mg/dose) PO BID x2 (load), then 6.5 mg/kg PO BD (max: 200 mg/dose). Certain guidelines recommend initiation of HCQ for all hospitalized patients with COVID-19. ***Tocilizumab: Dose 8–12 mg/kg single dose. Interleukin-6 inhibitor, useful for children with rapid deterioration due to COVID-19. ****Remdesivir: Investigational antiviral drug for COVID-19. *****Azithromycin: Dose 10 mg/kg/day for 1 day, then 5 mg/kg/day for 4 days. ******Lopinavir-Ritonavir: Dose 15–25 kg: 200 mg–50 mg; 26–35 kg: 300 mg–75 mg; >35 kg: 400 mg–100 mg PO BID for 5 days

vertical transmission of infection from mother to foetus occurs in the intrauterine environment (Lu and Shi 2020; Chen et al. 2020b). Early isolation is also recommended for children with underlying disease manifestations (Yang et al. 2020).

7.6 Conclusions

Despite low mortality and low infection rate among children and adolescents, they play a crucial role in the spread of infection in this ongoing pandemic of coronavirus disease. Adequate prevention measures, early identification and isolation will be helpful in altering the course of this pandemic.

7.7 Future Perspectives

There is a paucity of research on coronavirus infection in children and adolescents. There is a need to monitor the long-term impact of the virus exposure on the growth, development and other health measures. There is also an intense need to explore treatment options and vaccination for the effective control of coronavirus infection.

Executive Summary

- Children and adolescents with coronavirus infection have milder symptoms.
- Compromised immune function, malnutrition, co-morbid medical illnesses and poor hygiene are potential risk factors for contracting coronavirus infection in children and adolescents.
- Social distancing, limiting group activities, play time, tours, picnics and adequate hygiene training may be beneficial in limiting the chances of getting coronavirus infection in children and adolescents.

References

American Academy of Pediatrics (2020) COVID-19 in children: initial characterization of the pediatric disease. Pediatrics 145(4):e20200834

Centers for Disease Control and Prevention (2020) Coronavirus disease 2019 (COVID-19). Centers for Disease Control and Prevention. [cited 2020 Mar 22]. https://www.cdc.gov/Coronavirus/2019-ncov/faq.html?CDC_AA_refVal=https%3A%2F%2Fwww.cdc.gov%2FCoronavirus%2F2019-ncov%2Fprepare%2Fchildren-faq.html#school-dismissals

Chen Z-M, Fu J-F, Shu Q, Chen Y-H, Hua C-Z, Li F-B et al (2020a) Diagnosis and treatment recommendations for pediatric respiratory infection caused by the 2019 novel coronavirus. World J Pediatr. 10.1007/s12519-020-00345-5. Advance online publication. https://doi.org/10.1007/s12519-020-00345-5

Chen H, Guo J, Wang C, Luo F, Yu X, Zhang W et al (2020b) Clinical characteristics and intrauterine vertical transmission potential of COVID-19 infection in nine pregnant women: a retrospective review of medical records. Lancet 395(10226):809–815. [cited 2020 Mar 23]. http://www.sciencedirect.com/science/article/pii/S0140673620303603

Kam K, Yung CF, Cui L, Lin Tzer Pin R, Mak TM, Maiwald M et al (2020) A well infant with coronavirus disease 2019 (COVID-19) with high viral load. Clin Infect Dis. ciaa201. Advance online publication. https://doi.org/10.1093/cid/ciaa201. [cited 2020 Mar 23]. https://academic.oup.com/cid/advance-article/doi/10.1093/cid/ciaa201/5766416

Kar SK, Tripathy S (2019) Risk factors. In: Shackelford TK, Weekes-Shackelford VA (eds) Encyclopedia of evolutionary psychological science. Springer International Publishing, Cham, pp 1–4. https://doi.org/10.1007/978-3-319-16999-6_800-1

Liu W, Zhang Q, Chen J, Xiang R, Song H, Shu S et al. (2020) Detection of COVID-19 in children in early January 2020 in Wuhan, China. N Engl J Med 382(14):1370–1371 [cited 2020 Mar 23]. https://doi.org/10.1056/NEJMc2003717

Lu Q, Shi Y (2020) Coronavirus disease (COVID-19) and neonate: what neonatologist need to know. J Med Virol. 10.1002/jmv.25740. Advance online publication. https://doi.org/10.1002/jmv.25740. [cited 2020 Mar 23]. https://onlinelibrary.wiley.com/doi/abs/10.1002/jmv.25740

WHO (2020) Novel coronavirus (COVID-19) situation. [cited 2020 Mar 21]. https://experience.arcgis.com/experience/685d0ace521648f8a5beeeee1b9125cd

World Health Organization (2020) Report of the WHO-China Joint Mission on Coronavirus Disease 2019 (COVID-19). World Health Organization. [cited 2020 Mar 22]. https://www.who.int/docs/default-source/Coronaviruse/who-china-joint-mission-on-covid-19-final-report.pdf

Worldometers (2020) Age, sex, existing conditions of COVID-19 cases and deaths. February 29, 4:40 GMT [cited 2020 Mar 22]. https://www.worldometers.info/Coronavirus/Coronavirus-age-sex-demographics/#ref-2

Wu JT, Leung K, Bushman M, Kishore N, Niehus R, de Salazar PM et al (2020a) Estimating clinical severity of COVID-19 from the transmission dynamics in Wuhan, China. Nat Med 26 (4):506–510. https://doi.org/10.1038/s41591-020-0822-7. [cited 2020 Mar 21]. https://www.nature.com/articles/s41591-020-0822-7

Wu D, Wu T, Liu Q, Yang Z (2020b) The SARS-CoV-2 outbreak: what we know. Int J Infect Dis. Advance online publication. https://doi.org/10.1016/j.ijid.2020.03.004 [cited 2020 Mar 22]. https://www.ijidonline.com/article/S1201-9712(20)30123-5/abstract

Xia W, Shao J, Guo Y, Peng X, Li Z, Hu D (2020) Clinical and CT features in pediatric patients with COVID-19 infection: different points from adults. Pediatr Pulmonol 55(5):1169–1174. https://doi.org/10.1002/ppul.24718. [cited 2020 Mar 23]. https://onlinelibrary.wiley.com/doi/abs/10.1002/ppul.24718

Yang P, Liu P, Li D, Zhao D (2020) Corona virus disease 2019, a growing threat to children? J Infect. Advance online publication. https://doi.org/10.1016/j.jinf.2020.02.024

Zhang L, Liu Y (2020) Potential interventions for novel coronavirus in China: a systemic review. J Med Virol 92(5):479–490

Chapter 8
COVID-19: An Ophthalmological Update

Ankita, Apjit Kaur, and Shailendra K. Saxena

Abstract Ever since the newscast of the novel coronavirus outbreak in Wuhan and its subsequent spread to several countries worldwide, the possible modes of spread are being anticipated by various health care professionals. Tear and other conjunctival secretions, being one of the body fluids, can potentially help transmit the disease inadvertently. Conjunctival secretions from patients and asymptomatic contacts of COVID-19 cases may also spread the disease further into the community. Direct inoculation of body fluids into the conjunctiva of healthy individual is also postulated to be another mode of spread. The risk to heath care providers thus becomes strikingly high. A vigilant ophthalmologist can play a critical role in breaking the chain of transmission.

Keywords COVID-19 · Novel coronavirus · Conjunctivitis · SARS-CoV-2

Abbreviations

COVID-19	Coronavirus disease-19
SARS-CoV	Severe acute respiratory syndrome coronavirus
SARS-CoV-2	Severe acute respiratory syndrome coronavirus-2

Ankita and Shailendra K. Saxena contributed equally as first author.

Ankita · A. Kaur (✉)
Department of Ophthalmology, King George's Medical University (KGMU), Lucknow, India

S. K. Saxena
Centre for Advanced Research (CFAR)-Stem Cell/Cell Culture Unit, Faculty of Medicine, King George's Medical University (KGMU), Lucknow, India
e-mail: shailen@kgmcindia.edu

S. K. Saxena (ed.), *Coronavirus Disease 2019 (COVID-19)*, Medical Virology: from Pathogenesis to Disease Control, https://doi.org/10.1007/978-981-15-4814-7_8

8.1 Introduction

The rapid increase in the reporting of patients suffering from COVID-19 has brought about a major heath emergency worldwide. COVID-19 was declared a pandemic by the World Health Organization (WHO) on March 11, 2020 (WHO 2020a). According to the WHO pandemic phases, COVID-19 is currently in Phase 5, which signifies global human-to-human spread of the virus (WHO 2009).

Though the major symptoms are mainly respiratory in nature, ocular symptoms are generally overlooked by clinicians. The major routes of transmission are through respiratory droplets and direct contact (Lei et al. 2018; Minodier et al. 2015; Zumla et al. 2015; Otter et al. 2016). Gastrointestinal tract and body fluids, including tears, can have the SARS-CoV-2 and may participate in the active spread of COVID-19, though the exact role remains uncertain (WHO 2020b).

Li Wenliang, a Chinese ophthalmologist at Wuhan Central Hospital, Hubei, the whistleblower of the disease outbreak, allegedly contracted the disease while attending a glaucoma patient and later succumbed to the illness (Wenliang 2020). Thus, ophthalmologists, as health care professionals, are also exposed to the possible impacts of COVID-19.

8.2 Ophthalmological Manifestation of Viral Respiratory Diseases

8.2.1 Spectrum of Presentation

The conjunctival sac is anatomically connected to the nasal cavity through the nasolacrimal duct. Most of the upper respiratory tract infections harbor the organism in the nasal mucosa (Thomas 2020). Since most of the acute URTI have viral etiology, adenovirus being the commonest, the appearance of conjunctivitis in such cases is not uncommon (Epling 2010; Azari and Barney 2013). Conjunctivitis is speculated to be an allergic immune response to the virus (Solano and Czyz 2020).

Conjunctival congestion with watery discharge is more characteristic of a viral conjunctivitis (Rietveld et al. 2003). There is a wide spectrum of presentation of viral conjunctivitis ranging from pink eye to hemorrhagic conjunctivitis.

The disease readily spreads from person to person and has the possibility of becoming an epidemic. Although viral conjunctivitis is most often mild and goes away without treatment within 1–2 weeks, it may last 2 or more weeks if complications develop. Major complications associated are punctate keratitis with subepithelial infiltrates, bacterial superinfection, conjunctival scarring and symblepharon, severe dry eye, irregular astigmatism, corneal ulceration with persistent keratoconjunctivitis, corneal scarring, and chronic infection (Bialasiewicz 2007).

8.2.2 Previous Coronavirus Infections in the Eye

Severe acute respiratory syndrome (SARS), a global health crisis, in 2003, was a result of infection with coronavirus strain (SARS-CoV).

Middle East respiratory syndrome–related coronavirus (MERS-CoV), first identified in Saudi Arabia in 2012, was another coronavirus-borne disease (Centers for Disease Control and Prevention 2020a).

Most of the previously encountered coronavirus species (SARS-CoV and MERS-CoV infections) rarely cause ocular infections in humans (Li et al. 2006). However, few reports suggest the presence of conjunctivitis of varying severity in patients suffering with coronavirus infections in different parts of the world (Finger 2003).

The presence of coronavirus in tear and conjunctival scrapings was also reported in few studies.

8.2.3 Novel Coronavirus Infection (COVID-19) in the Eye

Following the recent novel coronavirus outbreak in China, the presence of novel coronavirus in tears and conjunctival secretions of patients with COVID-19 was evaluated. Though majority of patients who had been enrolled for the study did not present with conjunctivitis, one who did had the coronavirus in tear and conjunctival swab. Conjunctivitis reported was mild, with a watery discharge. The results concluded that tears and conjunctival secretions had coronavirus in patients with conjunctivitis, but was absent in those without conjunctivitis. This also raised an alarm among the treating ophthalmologists towards the possible spread of the virus through tears, along with other body fluids (Xia et al. 2020).

Another study in China reported conjunctivitis in 9 patients with COVID-19, out of 1099 (0.8%). Few reports also suggest that initial symptoms of COVID-19 was not conjunctivitis, but the possibility of virus spread through conjunctiva cannot be excluded (Wei-Jie et al. 2020; Wang et al. 2020; Huang et al. 2020). The replication of virus in conjunctival epithelium is also enigmatic.

In view of the presence of coronavirus in body fluids of patients, and SARS-CoV-2 being similar to SARS-CoV, the risk of transmission through conjunctival secretion and tear cannot be neglected.

8.3 Combating the Ocular Spread—Ophthalmologist's Role Play

According to the recent American Academy of Ophthalmology (AAO) guidelines, there are multiple reports which suggest that the novel coronavirus can cause conjunctivitis and can also be transmitted by aerosol contact with conjunctiva.

Patients with conjunctivitis may initially report to an ophthalmologist, possibly making the eye care physician to first suspect a case of COVID-19 (Lu et al. 2020).

8.3.1 When to Suspect COVID-19 Conjunctivitis

- Ophthalmological intervention should depend upon the WHO pandemic phase in the concerned region (WHO 2009).
- In regions where community-level outbreak of COVID-19 is confirmed, every case of viral conjunctivitis with/without systemic symptoms should be considered as a probable case of COVID-19.
- In regions with impending community outbreak, every case of viral conjunctivitis with/without systemic symptoms should be suspected as a case of COVID-19.
- Confirmation of travel and contact history becomes less relevant in both the aforementioned scenarios.
- In regions with local spread, history of travel/contact/symptoms suggestive of COVID-19 should be enquired and cases belonging to group 2, 3, and 4 should be evaluated further.

8.3.2 Diagnosing and Managing COVID-19 Conjunctivitis

In regions with WHO pandemic phase >2, ruling out common viral pathogens causing conjunctivitis becomes secondary. Excluding the novel pathogen in such scenarios is of primary importance as it can help in the detection of subclinical cases to limit further spread of the disease.

According to a recently published literature on COVID-19 conjunctivitis, tear sample and conjunctival swab are reported to be positive for the novel coronavirus. So tear sample and conjunctival secretions need to be evaluated for the presence of the virus. Collection should be done using disposable sampling swab from the conjunctival fornix, preferably lower. The sample should be stored in universal transport medium at 4 °C and sent for RT-PCR (real-time Polymerase Chain Reaction) assay. Two such samples should be taken within 2–3 days' interval and assay should be repeated.

The reported sensitivity and specificity of RT-PCR are 85–87% and 100%, respectively, for diagnosing previous SARS-CoV infection. Recent data on the sensitivity and specificity of RT-PCR for novel coronavirus infection is not available. The treatment of COVID-19 conjunctivitis is an ongoing research (National Institutes of Health 2020). It might have a benign course, as cases reported are mild, and may be treated as normal viral conjunctivitis treatment protocol as of now.

8.4 Risk to Ophthalmologists

The transmission of COVID-19 can occur through the mucous membranes, including the conjunctiva. Thus, the treating ophthalmologist may be equally or potentially exposed to the health hazard. Several case reports have documented the spread of COVID-19 to ophthalmologists during routine diagnosis and treatment (Wang et al. 2020; Dai 2020; South China Morning Post 2020). Also, transmission of the disease through asymptomatic contacts has also been documented (Rothe et al. 2020).

Most of the ophthalmological diagnostic procedures (slit lamp examination, direct and indirect fundus examinations, tonometry, etc.) require close proximity with patients, thereby increasing the risk of exposure (Xia et al. 2020). Suboptimal infection control strategies at the health care level may unintentionally spread the disease and risk the entire community (Chan et al. 2006).

8.5 Preventive Measures in Ophthalmic Practice

According to the WHO cases classification schemes, patients can be grouped on the basis of triage system into general, suspect, and probable categories (World Health Organization 2003; Gavidia 2020). All patients should not attend the outpatient clinic to avoid personal and community spread of the disease. Triage should be performed and patients should be screened on the basis of:

- Travel history in the last 2–3 weeks to any of the hot spots of COVID infections (China, South Korea, Italy, Iran, etc.)
- History of contact with known COVID-19 patient/suspect
- Symptoms of cough, cold, or fever

Patients are categorized into four groups according to the triage system (Fig. 8.1). Groups 1 and 2 can be categorized as general patients and can be seen on an outpatient basis, but with personal protective equipment (PPE). Groups 3 and 4 should be characterized as suspect/probable cases and only cases having ophthalmic urgency should be managed, in isolation ward with full protection. Rest non-urgent ophthalmic appointments should be rescheduled after 2 weeks (Fig. 8.2). Figure 8.2 highlights the protocol for attending patients visiting ophthalmology clinics during COVID-19 outbreak (Huang et al. 2020; Rohit and Santosh 2020).

8.5.1 Outpatient Care

As per recent AAO guidelines, as a response to the state of national emergency due to COVID-19, eye care practitioners should reduce the number of outpatient

Groups	Patient presentation
1	Otherwise healthy patients with no travel/contact history in the last 2-3 weeks.
2	Patients appearing healthy but with recent contact/travel history, quarantined and declared unaffected.
3	Patients appearing healthy but with recent contact/travel history, not quarantined.
4	Patients with obvious signs of respiratory illness (cough, fever)

Fig. 8.1 Triage of patients presenting to ophthalmology clinic

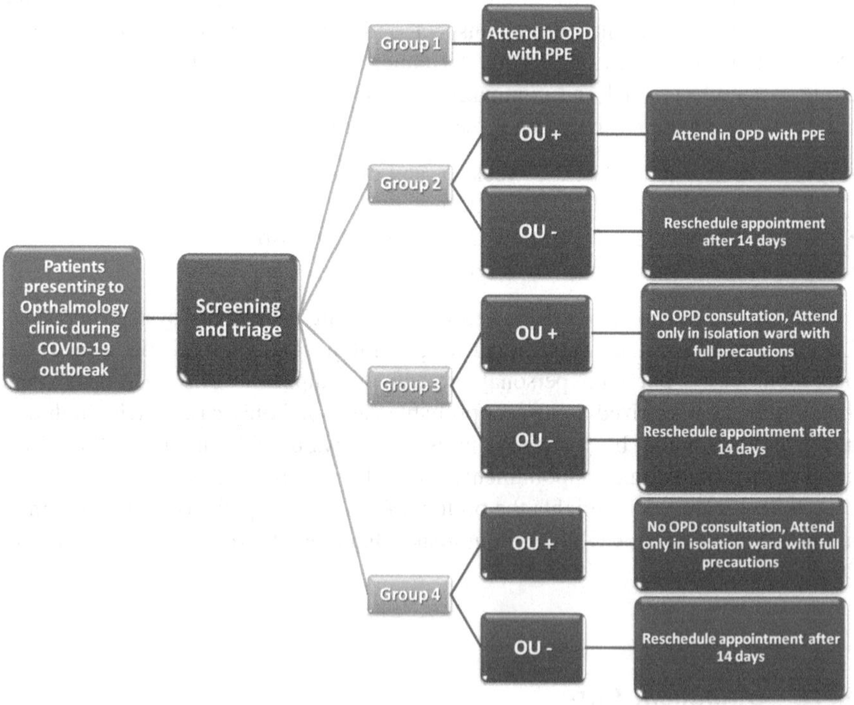

Fig. 8.2 Flow chart of attending patients visiting ophthalmology clinic. *OPD* outpatient department, *PPE* personal protective equipment, *OU* ophthalmic urgency

consultation days and elective surgical procedures, particularly in elderly patients and those with medical comorbidities (American Academy of Ophthalmology 2020).

The following preventive measures should be taken by the ophthalmologist in the outpatient clinic for general patients (Group 1 and 2) during such outbreaks:

8.5.1.1 Personal protection

Personal protective equipment (PPE) (Table 8.1) and general precautionary measures to be taken for outpatient procedures include

- Universal precaution:
  - Surgical/N95 respirator mask to cover the nose and mouth.
  - Gloves whenever body fluids (blood, secretions, urine, stool) are to be handled.
  - Water-repellant or water-resistant gowns.
  - Eye protective wear or goggles (visor).
- Hand sanitization:
  - Repeated hand washing, using hand rub (0.5–1.0% chlorhexidine in 80% ethyl alcohol).
- Gloves to be changed and hand hygiene practiced between each contact with different patients.
- Avoidance of touching of face shields, mask, eye protective wear, face, head, and neck area before thorough hand wash.

Table 8.1 Personal protective instrument to be used by health care practitioners for attending general and high-risk patients of COVID-19

Disposable wear/ protective measures	General patients (for outpatient care) (Group 1)	High-risk patients (for outpatient care and isolation wards) (Group 2, 3, and 4)
Cap	Standard	Standard
Eye protective wear	Visor or goggles	Face shield
Mask	Surgical mask	N95 respirator
Gown	Water repellent/water-resistant gowns	Water repellent/Barrier® surgical gown
Hand hygiene	Hand washing or alcohol-rub between patients	Glove
Shoe cover	Standard	Standard

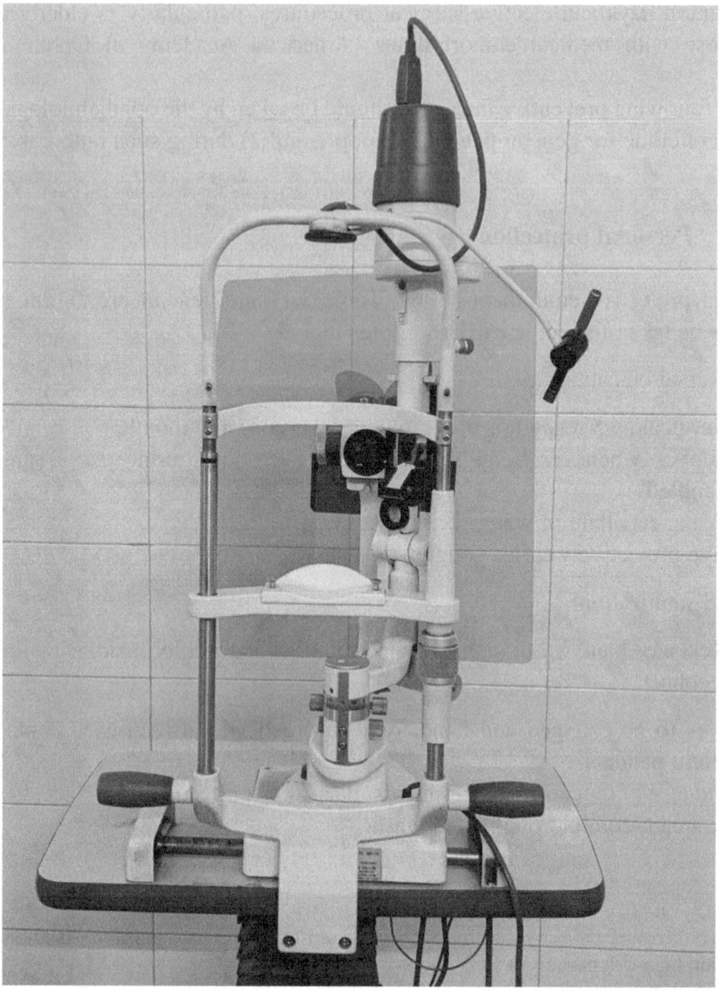

Fig. 8.3 Custom-made slit lamp breath shields

8.5.1.2 Examination Equipment Sterilization

- Slit lamps should accompany barriers or breath shields (Fig. 8.3).
- Applanation tonometry to be performed using Tonopen with a disposable sleeve. For Goldmann applanation tonometry, the prism tip must be sterilized between cases with alcohol swabs or bleach solution 1:10 as recommended. The prism tip should be immersed in bleach for at least 15 min.
- Ultrasonography probe to be sterilized with alcohol swab after every use.
- Scleral indentation for indirect ophthalmoscopy should be performed with disposable cotton swab stick.

- Direct ophthalmoscopy should be avoided, instead one should use slit lamp with breath shields and a non-contact lens for fundus evaluation.
- All ophthalmic instruments should be disinfected using diluted household bleach, alcohol solutions containing minimum of 70% alcohol, common EPA-registered household disinfectants including Clorox brand products, Lysol brand products, and Purell professional surface disinfectant wipes.

8.5.1.3 Consultation Room Hygiene and Ventilation

- Keep doors open in the clinic rooms to avoid handling of the door knobs and to maintain ventilation.
- Restriction of visitors in hospital premises unless under exceptional circumstances such as disabled patients or children.

For a patient who is suspected (or probable) to have COVID-19, the following measures should be taken in treatment by the ophthalmologist:

- For known cases of COVID-19, attending the patient on an outpatient basis is contraindicated due to high risk of cross infection. All consultations should be done inside quarantine or isolation wards.
- Any ophthalmic consultation should be deferred in a case suspected to have COVID-19, till the infective status of the patient is confirmed.

8.5.2 Inpatient Care

- For attending patients in isolation ward, personal protective equipment should be used as advised for high-risk cases.
- Disposable protective wear should be discarded separately, without touching any part of the skin/face.

8.5.3 Operation Theater (OT)

- Only urgent ophthalmic procedures like acute angle closure glaucoma, rhegmatogenous retinal detachment with macula on, traumatic rupture of the eye, and intraocular foreign body to be operated (South China Morning Post 2020).
- All other elective surgeries to be deferred till crisis subsides.
- Operating surgeon to be prepared with full precautions—SMS surgical gowns, shoe legging, face mask, surgeon cap, gloves, N95 mask.

- Removal of gown—special care to be taken in order to prevent outer surface from touching any part of the skin.
- Separate collection bag for waste generated and disposables used during surgery.
- High-level disinfection practice for cleaning of operating room (Rutala et al. 1993; Centers for Disease Control and Prevention 2020b).
  - Laminar air flow to be maintained in the room.
  - Operating microscope—disinfection with 2% glutaraldehyde.
  - Instrument sterilization—disinfection with 2% glutaraldehyde, sterilization using ethylene oxide (ETO) for moisture and heat-sensitive instruments, hydrogen peroxide plasma sterilization for all other surgical instruments.
  - Operation room fumigation with 5% formaldehyde.
- No noninfected cases to be operated in the same OT on the same day under any circumstances.

8.5.4 Management in Case of Accidental Exposure

In case of accidental exposure to any body fluid of the infected patient, eyes should be thoroughly washed with running water and the skin area should be properly cleaned with at least 70% ethyl alcohol-based cleaning solution for at least 20 s. The person should be kept under quarantine till 2 weeks (Centers for Disease Control and Prevention 2020c). Meanwhile, nasopharyngeal sampling should be done regardless of the symptoms because developing the disease is almost inevitable in such cases (Centers for Disease Control and Prevention 2020c; Lauer et al. 2020). Symptomatic and supportive treatment should be given in those developing COVID-19. Postexposure prophylaxis is not yet available.

8.5.5 Economically Weaker Countries

Aforementioned precautions should be practiced in such places also. But due to the paucity of resources, the demands are mostly unmet.

In such scenarios, OPD consultation of any ophthalmological complaint should be withheld till the status of infection turns out negative. Isolation ward, however, if urgent, should be attended with full precautions. Any alternative to the standard precautionary measures may not be helpful in preventing infections. Self-quarantine should be done by the suspect patient or health care provider for at least 2 weeks duration.

8.6 Conclusions

Ophthalmologists can play a key role in outbreak detection, surveillance, and early response to COVID-19. Unawareness at the eye care level may result in mass infections among health care providers and patients.

8.7 Future Perspectives

The outbreak of COVID-19 has brought out the stark inadequacies of the health care system at various levels, specially the primary and secondary care. It is high time the policy makers should take initiatives to strengthen primary and secondary heath care and increase investment in health, including ophthalmic care sector.

Previous studies had warned about a possible severe coronavirus pandemic, similar to SARS and MERS in near future. In order to fight forthcoming similar illnesses, proper sanitization and disinfection practices should be incorporated on a daily basis, by both doctors and people in the community. Situational awareness and adequate training should be give to health care providers of all specialties to combat outbreak of new infectious diseases. Early preventive measures should be undertaken to prevent pandemics. Cooperation among both public and private hospital setups is crucial for management.

Highlights

- Early identification of COVID-19 can be done by a vigilant ophthalmologist, reducing risk of further human-to-human transmission.
- Conjunctivitis can be the first symptom of COVID-19 in patients with positive travel/contact history.
- SARS-CoV-2 is present in patients with COVID-19 conjunctivitis.
- High risk of transmission is present through direct inoculation into conjunctiva.
- Patients having no respiratory symptoms except COVID-19 conjunctivitis are also infectious.

Acknowledgments The authors are grateful to the Vice Chancellor, King George's Medical University (KGMU), Lucknow, India, for the encouragement of this work. The authors have no other relevant affiliations or financial involvement with any organization or entity with a financial interest in or financial conflict with the subject matter or materials discussed in the manuscript apart from those disclosed.

Conflicts of interest: None.

References

American Academy of Ophthalmology (2020) Alert: important coronavirus updates for ophthalmologists. American Academy of Ophthalmology. [cited 2020 Mar 19]. https://www.aao.org/headline/alert-important-coronavirus-context

Azari A, Barney N (2013) Conjunctivitis. JAMA 310(16):1721–1729

Bialasiewicz A (2007) Adenoviral keratoconjunctivitis. Sultan Qaboos Univ Med J 7(1):15–23. [cited 2020Mar19]. https://www.ncbi.nlm.nih.gov/pmc/articles/PMC3086413/

Centers for Disease Control and Prevention (2020a) About MERS. Centers for Disease Control and Prevention. [cited 11 March 2020]. https://www.cdc.gov/coronavirus/mers/about/index.html

Centers for Disease Control and Prevention (2020b) Division of Healthcare Quality Promotion (DHQP). Centers for Disease Control and Prevention. [cited 2020 Mar 19]. https://www.cdc.gov/ncezid/dhqp/index.html

Centers for Disease Control and Prevention (2020c) Clinical specimens: novel coronavirus (2019-nCoV). Centers for Disease Control and Prevention. 2020 [cited 2020 Mar 19]. https://www.cdc.gov/coronavirus/2019-nCoV/lab/guidelines-clinical-specimens.html

Chan W-M, Liu DTL, Chan PKS, Chong KKL, Yuen KSC, Chiu TYH et al (2006) Precautions in ophthalmic practice in a hospital with a major acute SARS outbreak: an experience from Hong Kong. Eye 20:283–289. https://doi.org/10.1038/sj.eye.6701885

Dai X (2020) Peking University Hospital Wang Guangfa disclosed treatment status on Weibo and suspected infection without wearing goggles. http://www.bjnews.com.cn/news/2020/01/23/678189.html. Accessed 23 Jan 2020

Epling J (2010) Bacterial conjunctivitis. BMJ Clin Evid. 0704:21. https://www.ncbi.nlm.nih.gov/pmc/articles/PMC2907624/#

Finger C (2003) Brazil faces worst outbreak of conjunctivitis in 20 years. Lancet 361(9370):1714. [cited 2020 Mar 19]. https://www.ncbi.nlm.nih.gov/pubmed/12767749

Gavidia M (2020) American Academy of Ophthalmology issues coronavirus guidelines. AJMC. [cited 2020 Mar 11]. https://www.ajmc.com/newsroom/american-academy-of-ophthalmology-issues-coronavirus-guidelines

Huang C, Wang Y, Li X, Ren L, Zhao J, Hu Y, Zhang L et al (2020) Clinical features of patients infected with 2019 novel coronavirus in Wuhan, China. Lancet 395(10223):497–506. https://doi.org/10.1016/S0140-6736(20)30183-5

Lauer SA, Grantz KH, Bi Q, Jones FK, Zheng Q, Meredith HR, et al (2020) The incubation period of coronavirus disease 2019 (COVID-19) from publicly reported confirmed cases: estimation and application. Ann Intern Med. [cited 2020 Mar 19]. https://annals.org/aim/fullarticle/2762808/incubation-period-coronavirus-disease-2019-covid-19-from-publicly-reported

Lei H, Li Y, Xiao S, Lin C, Norris S, Wei D et al (2018) Routes of transmission of influenza A H1N1, SARS CoV, and norovirus in air cabin: comparative analyses. Indoor Air 28(3):394–403

Li W, Wong S, Li F, Kuhn J, Huang I, Choe H et al (2006) Animal origins of the severe acute respiratory syndrome coronavirus: insight from ACE2-S-protein interactions. J Virol 80(9):4211–4219. https://doi.org/10.1128/JVI.80.9.4211-4219.2006

Lu C, Liu X, Jia Z (2020) 2019-nCoV transmission through the ocular surface must not be ignored. Lancet 395(10224):39. https://doi.org/10.1016/S0140-6736(20)30313-5

Minodier L, Charrel R, Ceccaldi P, van der Werf S, Blanchon T, Hanslik T et al (2015) Prevalence of gastrointestinal symptoms in patients with influenza, clinical significance, and pathophysiology of human influenza viruses in faecal samples: what do we know? Virol J 12(1)

National Institutes of Health (2020) Coronavirus disease 2019 (COVID-19). U.S. Department of Health and Human Services. [cited 2020 Mar 19]. https://www.nih.gov/health-information/coronavirus

Otter J, Donskey C, Yezli S, Douthwaite S, Goldenberg S, Weber D (2016) Transmission of SARS and MERS coronaviruses and influenza virus in healthcare settings: the possible role of dry surface contamination. J Hosp Infect 92(3):235–250

Rietveld R, Van Weert H, ter Riet G, Bindels P (2003) Diagnostic impact of signs and symptoms in acute infectious conjunctivitis. BMJ 327(7418):789

Rohit CK, Santosh GH (2020) All eyes on coronavirus—what do we need to know as ophthalmologists. Indian J Ophthalmol 68:549–553

Rothe C, Schunk M, Sothmann P, Bretzel G, Froeschl G, Wallrauch C et al (2020) Transmission of 2019-nCoV infection from an asymptomatic contact in Germany. N Engl J Med 382 (10):970–971. https://doi.org/10.1056/NEJMc2001468. Epub 2020 Jan 30

Rutala WA, Gergen MF, Weber DJ (1993) Evaluation of a rapid readout biological indicator for flash sterilization with three biological indicators and three chemical indicators. Infect Control Hosp Epidemiol 14(7):390–394. U.S. National Library of Medicine. [cited 2020 Mar 19]. https://www.ncbi.nlm.nih.gov/pubmed/8354870

Solano D, Czyz CN (2020) Viral conjunctivitis. [Updated 2020 Feb 5]. In: StatPearls. StatPearls Publishing, Treasure Island, FL. https://www.ncbi.nlm.nih.gov/books/NBK470271/#

South China Morning Post (2020) Chinese expert thinks he contracted coronavirus through his eyeballs. South China Morning Post. [cited 2020 Mar 19]. https://www.scmp.com/news/china/article/3047394/chinese-expert-who-came-down-wuhan-coronavirus-after-saying-it-was

Thomas M (2020) Upper respiratory tract infection. StatPearls. U.S. National Library of Medicine. [cited 2020 Mar19]. https://www.ncbi.nlm.nih.gov/books/NBK532961/

Wang D, Hu B, Hu C, Fangfang Z, Xing L, Jing Z et al (2020) Clinical characteristics of 138 hospitalized patients with 2019 novel coronavirus–infected pneumonia in Wuhan, China. JAMA 323(11):1061–1069. https://doi.org/10.1001/jama.2020.1585

Wei-Jie G, Zheng-Yi N, Yu H, Wen L, Chun-Quan O, Jian-Xing H et al (2020) Clinical characteristics of coronavirus disease 2019 in China. N Engl J Med. https://doi.org/10.1056/NEJMoa2002032

Wenliang L (2020) En.wikipedia.org. [cited 19 March 2020]. https://en.wikipedia.org/wiki/Li_Wenliang

WHO (2009) Pandemic influenza preparedness and response: a WHO guidance document. World Health Organization, Geneva. 4, The WHO pandemic phases. https://www.ncbi.nlm.nih.gov/books/NBK143061/

WHO (2020a) Director-General's opening remarks at the media briefing on COVID-19, 11 March 2020. Who.int. 2020 [cited 19 March 2020]. https://www.who.int/dg/speeches/detail/who-director-general-s-opening-remarks-at-the-media-briefing-on-covid-19%2D%2D-11-march-2020

WHO (2020b) Update 27—one month into the global SARS outbreak: status of the outbreak and lessons for the immediate future. Who.int. 2020 [cited 19 March 2020]. https://www.who.int/csr/sars/archive/2003_04_11/en/

World Health Organization (2003) Case definitions for surveillance of severe acute respiratory syndrome (SARS) May 1, 2003. http://www.who.int/csr/sars/casedefinition

Xia J, Tong J, Liu M, Shen Y, Guo D (2020) Evaluation of coronavirus in tears and conjunctival secretions of patients with SARS-CoV-2 infection. J Med Virol. https://doi.org/10.1002/jmv.25725

Zumla A, Hui D, Perlman S (2015) Middle East respiratory syndrome. Lancet 386(9997):995–1007

Chapter 9
Laboratory Diagnosis of Novel Coronavirus Disease 2019 (COVID-19) Infection

Abhishek Padhi, Swatantra Kumar, Ekta Gupta, and Shailendra K. Saxena

Abstract Coronavirus disease 2019 (COVID-19) is an infection caused by the novel coronavirus severe acute respiratory coronavirus 2 (SARS-CoV-2). The infection manifests as a mild flu to severe acute respiratory infection. The World Health Organization (WHO) declared COVID-19 as a global pandemic on March 11, 2020. The disease spreads by droplet infection from person to person. Early diagnosis is the key for prompt management of cases and control of the spread of the virus. Currently, the laboratory diagnosis of SARS-CoV-2 is based on nucleic acid amplification tests (NAAT) like real-time reverse transcriptase (RT-PCR). Various genes like E, N, S, ORF and RdRp are targeted as a part of screening and confirmation of cases. Furthermore, nucleic acid sequencing may be done for the identification of mutation in the genome of SARS-CoV-2. The development of serological assays and point of care molecular test will further intensify the diagnostic modalities of SARS-CoV-2.

Keywords COVID-19 · SARS-CoV-2 · Laboratory diagnosis · Molecular method

9.1 Introduction

Coronavirus disease 2019 (COVID-19) is a severe acute respiratory infection caused by the novel coronavirus severe acute respiratory syndrome coronavirus 2 (SARS-CoV-2) (WHO 2020a), which started initially as a cluster of cases from Wuhan,

---

Abhishek Padhi and Shailendra K. Saxena contributed equally as first author.

A. Padhi · E. Gupta (✉)
Department of Clinical Virology, Institute of Liver and Biliary Sciences, New Delhi, India

S. Kumar · S. K. Saxena
Centre for Advanced Research (CFAR), Faculty of Medicine, King George's Medical University (KGMU), Lucknow, India
e-mail: shailen@kgmcindia.edu

S. K. Saxena (ed.), *Coronavirus Disease 2019 (COVID-19)*, Medical Virology: from Pathogenesis to Disease Control, https://doi.org/10.1007/978-981-15-4814-7_9

China (Zhu et al. 2020), has now spread to 135 countries with 142,539 number of confirmed cases and 5393 deaths (WHO 2020b). The World Health Organization declared COVID-19 as a global pandemic on March 11, 2020 (WHO 2020c). The disease primarily spreads via close contact of respiratory droplets generated by infected individuals (Center for Disease Control and Prevention 2020a). At the global level, sufficient testing capacity for COVID-19 is not available as it should be and therefore preventing individuals from accessing care. During the initial outbreak period, different countries have followed and implemented various testing strategies, depending on the availability of diagnostics and consumables. However, strict steps taken by the WHO has made the diagnostic available with the mission to "detect, protect and treat" to break the chain of transmission of SARS-CoV-2 (WHO 2020d). Therefore, early diagnosis and prompt treatment can substantially reduce the number of prospective cases. Hence, laboratory diagnosis of SARS-CoV-2 holds the key in containing and restricting the COVID-19 pandemic.

People who are in close contact with suspicious exposure have been advised under a 14-day health observation period that should be started from the last day of contact with infected individuals. Once these individuals show any symptoms including coughing, sneezing, shortness of breath or diarrhoea, they should require immediate medical attention. Immediate isolation of the suspected individual should be performed with proper guidelines, and they should be closely monitored for clinical symptoms and diagnosis should be performed in hospital-based laboratories as soon as possible. In addition, surveillance should be performed for those who were in contact with the suspected or conformed individuals by observing their clinical symptoms. Before taking decision about isolation, authorities should make sure that whether the suspected individual requires home isolation and careful clinical evaluation with safety assessment by healthcare professionals or not. If the suspected individuals present any symptoms during isolation, they should contact the doctors for their treatment. During home isolation, suggested medication and symptoms should be closely recorded. The suspected, probable and confirmed case definition of COVID-19 by the WHO has been presented in Fig. 9.1. The decision for diagnosis of an individual should be based on epidemiological and clinical factors which linked to an assessment of the probability of infection.

9.2 Specimen Types and Collection

Adequate standard operating procedures (SOPs) should be followed before collecting any specimen, including the accurate training of the staff for appropriate specimen collection, packaging, storage and transport. The staff should be very well aware of the preventive measures and control guidelines for COVID-19, for that purpose WHO interim guidance should be followed. Collected specimens should be regarded as potentially infectious and therefore extreme precautions should be taken during the handling of the samples. The diagnosis of clinical specimens collected from the suspected individuals should be executed in appropriately equipped

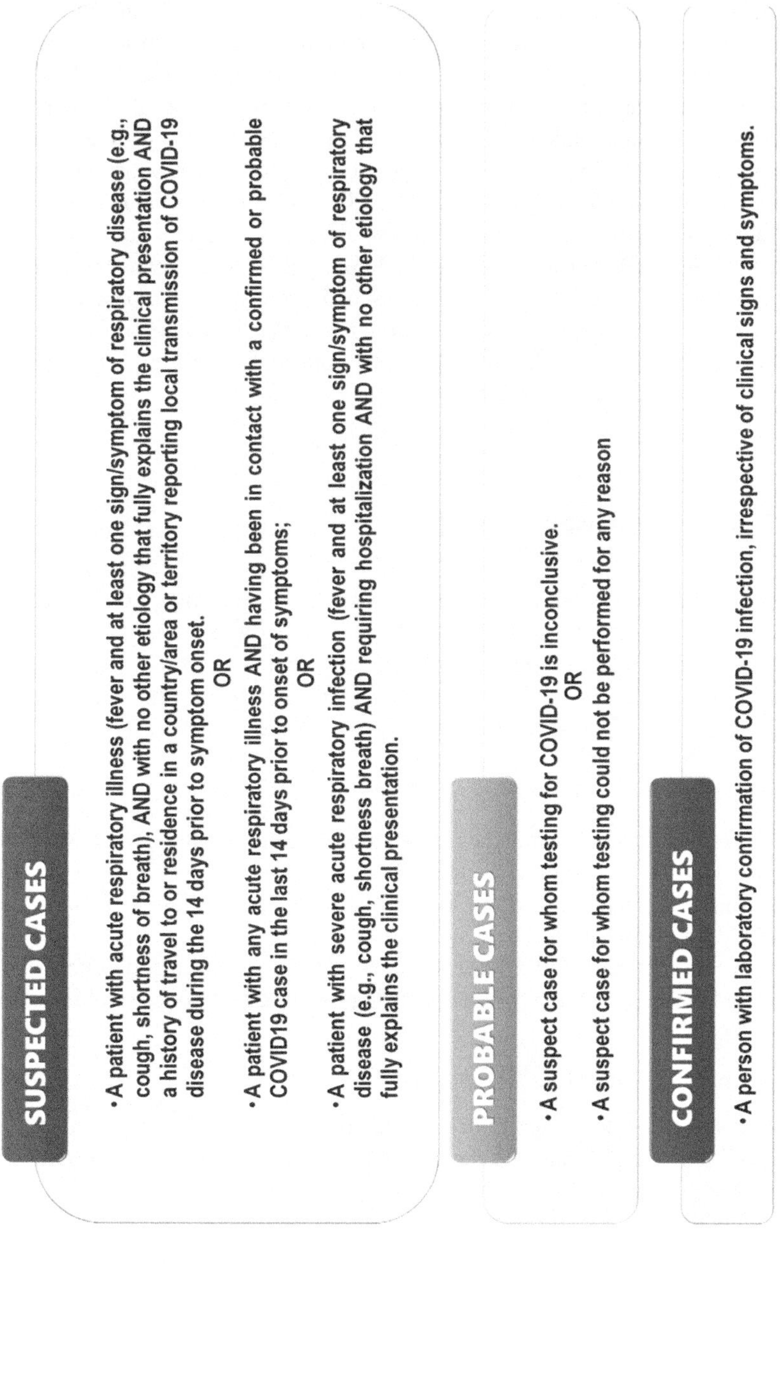
SUSPECTED CASES
• A patient with acute respiratory illness (fever and at least one sign/symptom of respiratory disease (e.g., cough, shortness of breath), AND with no other etiology that fully explains the clinical presentation AND a history of travel to or residence in a country/area or territory reporting local transmission of COVID-19 disease during the 14 days prior to symptom onset.
OR
• A patient with any acute respiratory illness AND having been in contact with a confirmed or probable COVID19 case in the last 14 days prior to onset of symptoms;
OR
• A patient with severe acute respiratory infection (fever and at least one sign/symptom of respiratory disease (e.g., cough, shortness breath) AND requiring hospitalization AND with no other etiology that fully explains the clinical presentation.
PROBABLE CASES
• A suspect case for whom testing for COVID-19 is inconclusive.
OR
• A suspect case for whom testing could not be performed for any reason
CONFIRMED CASES
• A person with laboratory confirmation of COVID-19 infection, irrespective of clinical signs and symptoms.

Fig. 9.1 Case definition of COVID-19 by the World Health Organization

laboratories by the staff specifically trained in technical and biosafety measures. National guidelines on laboratory biosafety should be strictly followed, and all the procedures should be undertaken based on a risk assessment. Specimens for molecular diagnostics require BSL-2 or equivalent facilities, whereas any attempt to culture virus requires BSL-3 facilities at minimum. At least respiratory material should be collected for the diagnosis of COVID-19 from suspected individuals. Upper respiratory tract specimens nasopharyngeal and oropharyngeal swab or wash in ambulatory patients including lower respiratory specimens such as sputum (if generated) and/or bronchoalveolar lavage or endotracheal aspirate in patients with more severe respiratory disease (WHO 2020e). Other clinical samples may also be collected as SARS-CoV-2 has also been found in blood and stool, similar to SARS and MERS (Wu et al. 2020). In addition, paired sera should be collected during acute and convalescent phase for retrospective study using serological assays (when available). For post-mortem study, lung tissue may be collected for studying the pathophysiology of the disease. Specimen collection details proposed by the WHO have been presented in Table 9.1.

9.3 Biosafety Measures

Samples should be collected by well-trained healthcare personnel, donning proper personal protective equipment (PPE) which includes masks, gloves, goggles, gown, head cover, shoe cover and hand sanitizer, soap and water following adequate infection control measures including hand hygiene and adequate biosafety precautions to protect individual and the environment (Center for Disease Control and Prevention 2020b).

The correct sequence of donning the PPE:

- Home clothes, jewellery, watches, rings, bangles, etc. should be removed and hospital scrub suite should be donned.
- Proper hand hygiene to be done using alcohol-based hand rub or soap and water prior to donning of the PPE.
- Sequence of donning PPE: Shoe cover → clean, disposable non-permeable gown → N95 respirator with proper fit testing → Eye goggles/face shield → Head cover → Gloves.

The correct sequence of doffing the PPE:

- Doffing should be done only in designated areas.
- Any soiling in the PPE must be checked before doffing. If any, the area should be disinfected before doffing.
- Hand hygiene must be followed after every step.
- Sequence of doffing the PPE: Shoe cover → Gloves → Eye goggles/face shield → Head cover → Gown → N95 respirator.

Table 9.1 Specimen collection details

Specimen type	Collection materials	Transport to laboratory	Storage temperature till testing	Comments
Nasopharyngeal and oropharyngeal swab[a]	Dacron or polyester flocked swabs[b]	4°C	≤ 5days : 4°C >5 days : -70°C	To increase the viral load both nasopharyngeal and oropharyngeal swabs should be placed in the same tube
Bronchoalveolar lavage	Sterile container[b]	4°C	≤ 48 hours : 4°C >48 hours : -70°C	Some dilution of pathogen may be there but a important specimen in patients with serious infection
Tracheal aspirate, nasopharyngeal aspirate or nasal wash	Sterile container[b]	4°C	≤ 48 hours : 4°C >48 hours : -70°C	-
Sputum	Sterile container	4°C	≤ 48 hours : 4°C >48 hours : -70°C	To ensure if the material is from lower respiratory tract
Tissue from biopsy or autopsy including from lung	Sterile container with saline or VTM	4°C	≤ 24 hours : 4°C >24 hours : -70°C	Important for post mortem diagnosis
Serum (acute and convalescent samples)[a]	Serum separator tubes (adults: collect 3-5 ml whole blood)	4°C	≤ 5days : 4°C >5 days : -70°C	Paired samples to be collected : Acute – first week of illness Chronic – 2 to 3 weeks later
Whole blood (5ml)	Blood in EDTA vial	4°C	≤ 5days : 4°C >5 days : -70°C	-
Stool	Stool container	4°C	≤ 5days : 4°C >5 days : -70°C	Important sample to rule out gastrointestinal infection
Urine	Urine collection container	4°C	≤ 5days : 4°C >5 days : -70°C	-

Adapted from the WHO guidelines
[a]Mandatory specimen, [b]for transport of samples, viral transport medium (VTM) is to be used containing antibiotics and antifungal

- All the PPE must be disinfected and discarded following the local biomedical waste management rules.

  Steps for processing of samples:

- Initial processing (before inactivation) of all specimens should take place in a biological safety cabinet (BSC) or primary containment device (Center for Disease Control and Prevention 2020c).
- Laboratory work involving non-propagative methods (e.g. sequencing, nucleic acid amplification test [NAAT]) should be conducted in a Biosafety Level 2 (BSL-2) facility (Center for Disease Control and Prevention 2020c).
- Laboratory work involving propagative methods (e.g. virus culture, isolation or neutralization assays) should be conducted at a containment laboratory with

inward directional airflow in a Biosafety Level-3 (BSL-3) facility (Center for Disease Control and Prevention 2020c).

- Disinfectants having action against enveloped viruses should be used (e.g. hypochlorite [bleach], alcohol, hydrogen peroxide, quaternary ammonium compounds and phenolic compounds) (Center for Disease Control and Prevention 2020c).
- Patient specimens from suspected or confirmed cases should be transported as UN3373, "Biological Substance Category B". Viral cultures or isolates should be transported as Category A, UN2814, "infectious substance, affecting humans" (Center for Disease Control and Prevention 2020c).

9.4 Specimen Collection Packaging and Transport

After the collection, specimens should be transported to the laboratory as soon as possible for the diagnosis of COVID-19. During the transportation of the specimens correct handling of the specimens is indispensable. Specimen transportation should be performed and shipped under the cold chain maintenance of 2–8 °C. If in case of delay in delivery of the specimen to the laboratory, the specimens should be transported in a viral transport medium. It can be also frozen to −20 °C or ideally −70 °C and shipped on dry ice if further delays are likely to expect. During transportation, it is crucial to avoid repeated thawing and freezing of specimens. Transportation within national borders should comply with applicable rules and regulations. In case of international transportation of specimens, the UN Model Regulations should be followed along with other regulations depending on the mode of transport. The requirements for specimen collection, packaging and transport have been presented in Fig. 9.2 and the procedure for specimen packaging and transport has been presented in Fig. 9.3 (Center for Disease Control and Prevention 2020d). After specimen collection and packaging, the well-packed specimen (triple packaging as shown in Fig. 9.4) is shipped at the earliest to the nearest testing centre maintaining proper cold chain (WHO 2020f).

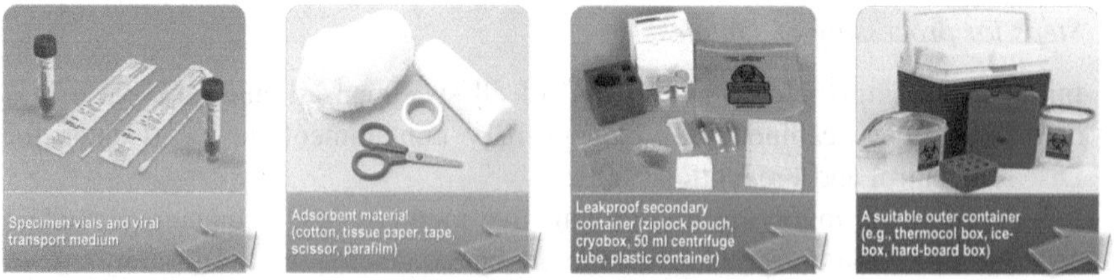

Fig. 9.2 Requirements for specimen collection, packaging and transport

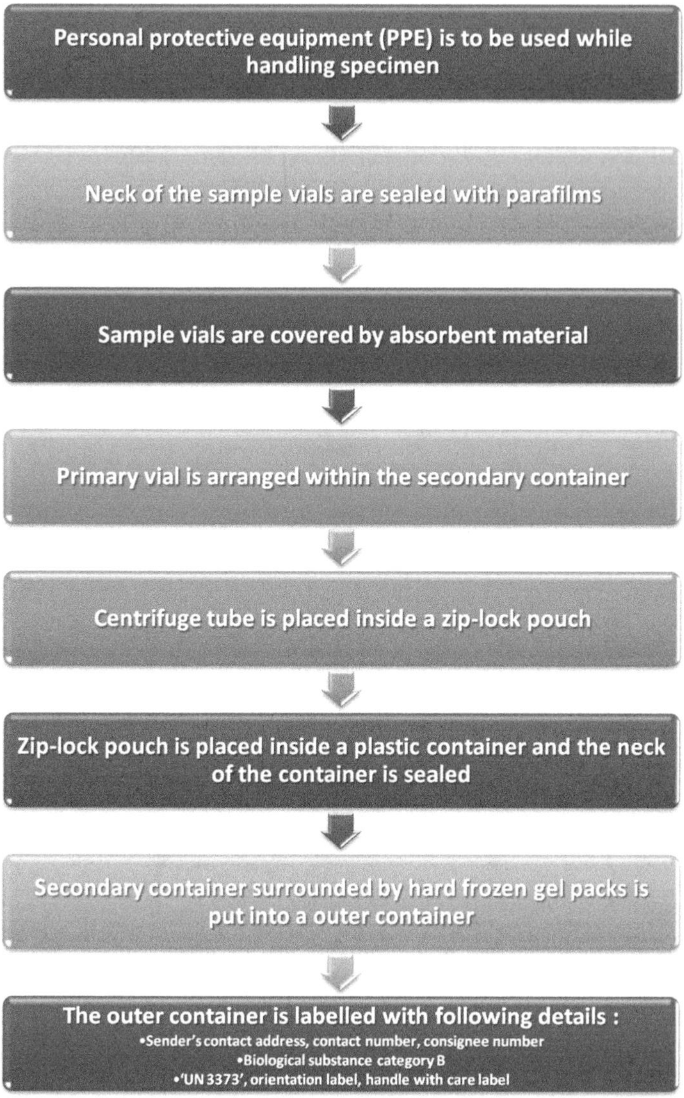

Fig. 9.3 Procedure for specimen packaging and transport

9.5 Diagnostic Methods for the Detection of SARS-CoV-2

9.5.1 Nucleic Acid Amplification Testing (NAAT)

At present confirmation of cases of COVID-19 is based on the detection of viral RNA by nucleic acid amplification tests (NAAT) such as real-time reverse transcriptase polymerase chain reactions (RT-PCR) with confirmation by nucleic acid

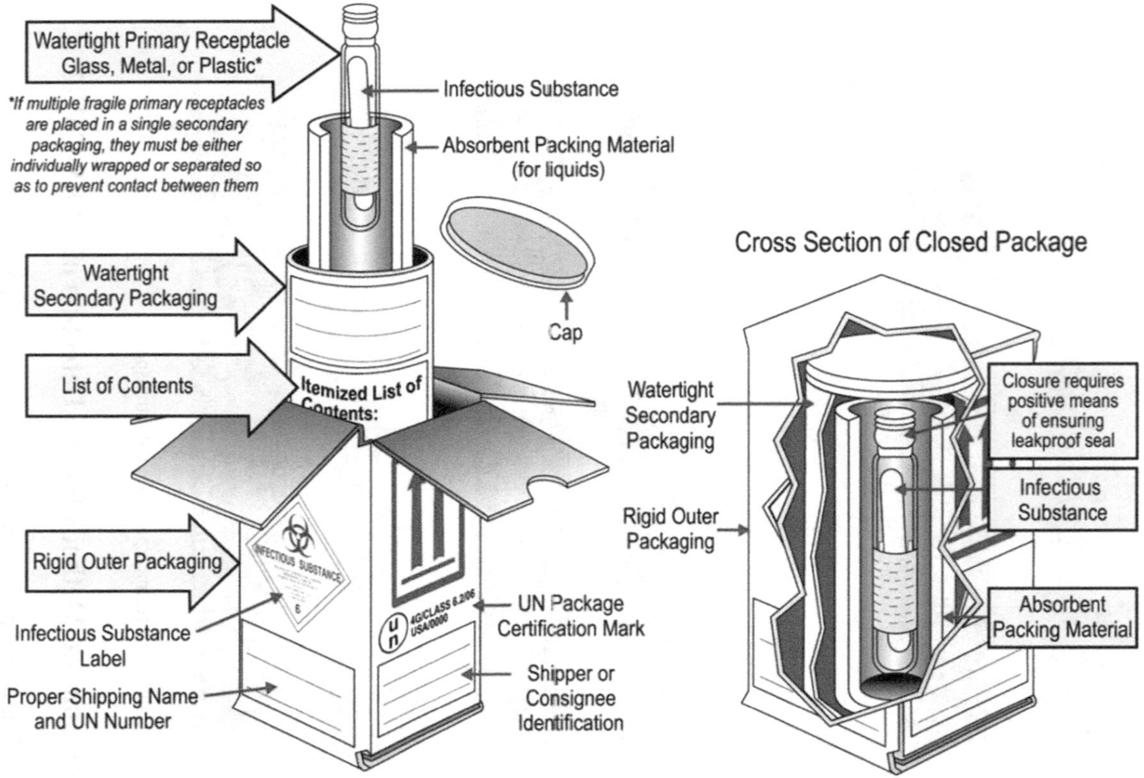

Fig. 9.4 Triple packaging of specimen. (Adapted from WHO (2020f))

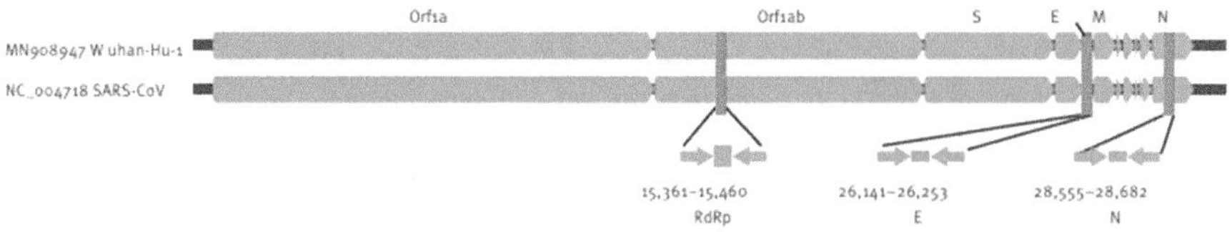

Fig. 9.5 Relative positions of amplicon targets on SARS-CoV-2 genome. *ORF* open reading frame, *RdRp* RNA-dependent RNA polymerase, *E* envelop protein gene, *N* nucleocapsid protein gene, *M* membrane protein gene, *S* spike protein gene

Table 9.2 Currently available protocol

Country	Institute	Gene targets
China	China CDC	ORF 1ab and N
Germany	Charitè	RdRp, E, N
Hong Kong SAR	HKU	ORF 1b-nsp14, N
Japan	National Institute of Infectious Diseases, Department of Virology	Pancorona and multiple targets, spike protein
Thailand	National Institute of Health	N
USA	US CDC	Three targets in N gene
France	Pasteur Institute, Paris	Two targets in RdRp gene

Adapted from WHO (2020e)

sequencing when necessary (WHO 2020e). The viral genes targeted so far include the N, E, S, ORF and RdRp genes (Fig. 9.5) according to SARS-CoV, GenBank NC_004718 (WHO 2020f). Different protocols followed by various countries are mentioned in Table 9.2.

One of the following conditions should be met to consider a case as laboratory confirmed by NAAT in areas with no SARS-CoV-2 circulation:

- A positive NAAT result for at least two different targets on the SARS-CoV-2 virus genome, of which at least one target is preferably specific for SARS-CoV-2 virus using a validated assay; OR
- One positive NAAT result for the presence of betacoronavirus, and SARS-CoV-2 virus further identified by sequencing partial or whole genome of the virus as long as the sequence target is larger or different from the amplicon probed in the NAAT assay used.

When there is ambiguity in results, sample should once again be collected from the patient and, if appropriate, sequencing of the virus from the original specimen or of an amplicon generated from an appropriate NAAT assay, different from the NAAT assay initially used, should be obtained to provide a reliable test result.

Areas where SARS-CoV-2 virus is widely circulating a simpler logarithm might suffice; for example screening of a single differential target is sufficient. One or more negative results do not rule out the possibility of SARS-CoV-2 virus infection. A number of factors could lead to a negative result in an infected individual, including:

- Poor quality of the specimen, containing little patient material (as a control, consider determining whether there is adequate human DNA in the sample by including a human target in the PCR testing).
- The specimen was collected late or very early in the infection.
- The specimen was not handled and shipped appropriately (non-maintenance of cold chain).
- Technical reasons inherent in the test, e.g. virus mutation or PCR inhibition.

If a negative result is obtained from a patient with a high index of suspicion for SARS-CoV-2 virus infection, particularly when only upper respiratory tract specimens were collected, additional specimens, including from the lower respiratory tract if possible, should be collected and tested.

9.5.2 Viral Sequencing

Sequencing does not have a role in the initial laboratory diagnosis of SARS-CoV-2 but can be helpful in the following circumstances:

- Providing confirmation of the presence of the virus.
- Monitor for viral genome mutations that might affect the performance of medical countermeasures, including diagnostic tests.
- Virus whole genome sequencing can also inform molecular epidemiology studies.

9.5.3 Serology

- Can aid investigation of an ongoing outbreak and retrospective assessment of the attack rate or extent of an outbreak.
- In cases where NAAT assays are negative and there is a strong epidemiological link to COVID-19 infection, paired serum samples (in the acute and convalescent phase) might support diagnosis once validated serology tests are available. Serum samples can be stored for these purposes.

9.5.4 Viral Culture

Viral culture is not recommended for the laboratory diagnosis of SARS-CoV-2. But viral culture can be used for research purposes like isolation of the virus, studying the properties of the virus and development of vaccine. Human airway epithelial cell lines were used for the initial isolation of the virus (Zhu et al. 2020).

9.6 Challenges for Diagnosis

Early diagnosis of COVID-19 is essential for the timely management as well as isolation of confirmed cases to prevent further transmission of patients. However, sample collection, transport and kit validation are major bottlenecks in the diagnosis of COVID-19. A study found that the total positivity of cases by initial RT-PCR were around 30–60% (Ai et al. 2020). This largely depends on the time at which sample has been collected as PCR positivity will be seen during the early days of symptoms. Furthermore, the sensitivity of the testing kits is a matter of debate and thereby a sizeable number of patients may not be identified, which may ultimately be detrimental in the early diagnosis and treatment of COVID-19 cases. Also in low- and middle-income countries (LMIC) (Hopman et al. 2020), the healthcare system is not robust enough as a result of which the testing laboratories often face difficulties in the performance of molecular testing.

Executive Summary

- *COVID-19*
  - Severe acute respiratory infection caused by novel coronavirus SARS-CoV-2.
  - First reported from Wuhan, China, as a cluster of cases with pneumonia.
  - The WHO declared COVID-19 as a global pandemic on March 11, 2019.
- *Sample types, collection and transport*
  - Vital samples: nasopharyngeal and oropharyngeal swabs collected in a single tube of viral transport media (VTM) to increase the yield of virus.
  - Other samples: paired sera (acute and convalescent phase serum sample), blood, bronchoalveolar lavage.
  - Sample packing to be done in a triple packing method and properly labelled.
  - Sample processing should be done in a biosafety level-2 facility.
- *Laboratory diagnosis*
  - Currently, laboratory diagnosis of SARS-CoV-2 is done by nucleic acid amplification testing like real-time reverse transcriptase PCR.
  - The viral genes targeted so far include the N, E, S, ORF and RdRp genes.
  - Various countries have submitted their primer probe designs to the WHO.
  - Many countries are adapting two-stage protocol.

(continued)

  - Example: for screening—E gene is targeted and for confirmation—RdRp gene is targeted (Germany protocol).
  - Sequencing partial or whole genome.
- *Challenges in diagnosis*
  - Sample collection, transport and kit performance are major bottlenecks in the diagnosis of COVID-19.
  - Availability of point of care tests which require minimal biosafety requirements.

9.7 Conclusions

Early diagnosis is the key for prompt management of COVID-19. Serological and molecular assays together will further strengthen the diagnosis of SARS-CoV-2. Laboratory networking is the need of the hour for real-time diagnosis of COVID-19.

9.8 Future Perspectives

Considering the above challenges for the diagnosis, a robust laboratory network is the need of the hour. India has a wide network of laboratories for testing of viral diseases; the Viral Research and Diagnostic Laboratories (VRDL) forms the base of the testing laboratories pyramid with an apex centre at the top of the pyramid. In these testing times, these VRDLs are taking the lead in the real-time diagnosis of COVID-19 in India. Such robust networking of laboratories can be replicated in LMICs for prompt sample collection till the final diagnosis of SARS-CoV-2 and other viral infections. Also the development of a serological assay at the earliest will be beneficial for resource-limited countries, and also the turnaround time will be significantly reduced. Furthermore, as the pandemic widens its arm a point of care molecular test will be like a holy grail in the rapid diagnosis of cases, thereby initiating the treatment at the earliest.

References

Ai T, Yang Z, Hou H, Zhan C, Chen C, Lv W, Tao Q, Sun Z, Xia L (2020) Correlation of chest CT and RT-PCR testing in coronavirus disease 2019 (COVID-19) in China: a report of 1014 cases. Radiology:200642. https://doi.org/10.1148/radiol.2020200642

Center for Disease Control and Prevention (2020a) Coronavirus disease 2019 (COVID-19) – transmission. In: Center for Disease Control and Prevention. https://www.cdc.gov/coronavirus/2019-ncov/prepare/transmission.html. Accessed 15 Mar 2020

Center for Disease Control and Prevention (2020b) Interim infection prevention and control recommendations for patients with suspected or confirmed coronavirus disease 2019 (COVID-19) in healthcare settings. Centre for Disease Control and Prevention. https://www.cdc.gov/coronavirus/2019-ncov/infection-control/control-recommendations.html. Accessed 25 Mar 2020

Center for Disease Control and Prevention (2020c) Interim Laboratory Biosafety Guidelines for handling and processing specimens associated with coronavirus disease 2019 (COVID-19). Centre for Disease Control and Prevention. https://www.cdc.gov/coronavirus/2019-ncov/lab/lab-biosafety-guidelines.html. Accessed 25 Mar 2020

Center for Disease Control and Prevention (2020d) Interim Guidelines for collecting, handling, and testing clinical specimens from persons for coronavirus disease 2019 (COVID-19). Centre for Disease Control and Prevention. https://www.cdc.gov/coronavirus/2019-ncov/lab/guidelines-clinical-specimens.html. Accessed 25 Mar 2020

Hopman J, Allegranzi B, Mehtar S (2020) Managing COVID-19 in low- and middle-income countries. JAMA. https://doi.org/10.1001/jama.2020.4169

WHO (2020a) Naming the coronavirus disease (COVID-19) and the virus that causes it. https://www.who.int/emergencies/diseases/novel-coronavirus-2019/technical-guidance/naming-the-coronavirus-disease-(covid-2019)-and-the-virus-that-causes-it. Accessed 15 Mar 2020

WHO (2020b) Novel coronavirus (2019-nCoV) situation reports. https://www.who.int/emergencies/diseases/novel-coronavirus-2019/situation-reports. Accessed 15 Mar 2020

WHO (2020c) WHO Director-General's opening remarks at the media briefing on COVID-19. https://www.who.int/dg/speeches/detail/who-director-general-s-opening-remarks-at-the-media-briefing-on-covid-19%2D%2D-11-march-2020. Accessed 15 Mar 2020

WHO (2020d) WHO Director-General's opening remarks at the media briefing on COVID-19. https://www.who.int/dg/speeches/detail/who-director-general-s-opening-remarks-at-the-mission-briefing-on-covid-19%2D%2D-13-march-2020. Accessed 15 Mar 2020

WHO (2020e) Laboratory testing for 2019 novel coronavirus (2019-nCoV) in suspected human cases. https://www.who.int/publications-detail/laboratory-testing-for-2019-novel-coronavirus-in-suspected-human-cases-20200117. Accessed 15 Mar 2020

WHO (2020f) Guidelines for the safe transport of infectious substances and diagnostic specimens. In: WHO. https://www.who.int/csr/resources/publications/biosafety/WHO_EMC_97_3_EN/en/. Accessed 18 Mar 2020

Wu Y, Guo C, Tang L, Hong Z, Zhou J, Dong X, Yin H, Xiao Q, Tang Y, Qu X, Kuang L, Fang X, Mishra N, Lu J, Shan H, Jiang G, Huang X (2020. pii: S2468-1253(20) 30083-2) Prolonged presence of SARS-CoV-2 viral RNA in faecal samples. Lancet Gastroenterol Hepatol. https://doi.org/10.1016/S2468-1253(20)30083-2

Zhu N, Zhang D, Wang W et al (2020) A novel coronavirus from patients with pneumonia in China, 2019. N Engl J Med 382:727–733

Chapter 10
Therapeutic Development and Drugs for the Treatment of COVID-19

Vimal K. Maurya, Swatantra Kumar, Madan L. B. Bhatt, and Shailendra K. Saxena

Abstract SARS-CoV-2/novel coronavirus (2019-nCoV) is a new strain that has recently been confirmed in Wuhan City, Hubei Province of China, and spreads to more than 165 countries of the world including India. The virus infection leads to 245,922 confirmed cases and 10,048 deaths worldwide as of March 20, 2020. Coronaviruses (CoVs) are lethal zoonotic viruses, highly pathogenic in nature, and responsible for diseases ranging from common cold to severe illness such as Middle East respiratory syndrome (MERS) and severe acute respiratory syndrome (SARS) in humans for the past 15 years. Considering the severity of the current and previous outbreaks, no approved antiviral agent or effective vaccines are present for the prevention and treatment of infection during the epidemics. Although, various molecules have been shown to be effective against coronaviruses both in vitro and in vivo, but the antiviral activities of these molecules are not well established in humans. Therefore, this chapter is planned to provide information about available treatment and preventive measures for the coronavirus infections during outbreaks. This chapter also discusses the possible role of supportive therapy, repurposing drugs, and complementary and alternative medicines for the management of coronaviruses including COVID-19.

Keywords SARS-CoV-2 · Novel coronavirus (2019-nCoV) · MERS-CoV · SARS-CoV · Complementary and alternative medicines · Repurposing drugs

---

V. K. Maurya · S. Kumar · M. L. B. Bhatt · S. K. Saxena (✉)
Department of Centre for Advanced Research (CFAR), Faculty of Medicine, King George's Medical University (KGMU), Lucknow, India
e-mail: shailen@kgmcindia.edu

S. K. Saxena (ed.), *Coronavirus Disease 2019 (COVID-19)*, Medical Virology: from Pathogenesis to Disease Control, https://doi.org/10.1007/978-981-15-4814-7_10

10.1 Introduction

Coronaviruses (CoVs) are enveloped, positive-sense RNA viruses belonging to the subfamily *Coronavirinae* and order *Nidovirales* (Huang et al. 2020). CoVs have a common genome organization and are commonly categorized as alpha CoVs, beta CoVs, gamma CoVs, and delta CoVs on the basis of genomic structures and phylogenetic relationships. In these CoVs, transmission of alpha CoVs and beta CoVs is restricted to mammals and induces respiratory illness in humans, while gamma CoVs and delta CoVs are known to affect birds and mammals (Song et al. 2019). Even though, most CoV infections are mild, but the outbreaks of the two beta CoVs, i.e., Middle East respiratory syndrome coronavirus (MERS-CoV) and severe acute respiratory syndrome coronavirus (SARS-CoV), have caused more than 10,000 combined cases and 1600 deaths in last 20 years (Skariyachan et al. 2019). The incubation period of MERS and SARS is given as 2–13 and 2–14 days, respectively, the progression of infection is rapid with MERS-CoV as compared to SARS-CoV, and the reported mortality rates were 34% and 10%, respectively (Chen et al. 2020; Rasmussen et al. 2016). CoVs are zoonotic in nature, which means these viruses are transmitted between animals and humans. In the same way, SARS and MARS were reported to transmit in humans from civet cats and camel, respectively (Shehata et al. 2016). Human-to-human transmission of SARS and MARS is also reported via close personal contact with infected patients. In 2012, MERS emerged as global heath challenge in countries near the Arabian Peninsula. As of July 31, 2019, 2458 laboratory-confirmed MERS cases and 848 deaths were reported where around 80% of these cases have been reported only in Saudi Arabia (Zheng et al. 2019). In 2003, SARS-CoV was originated in southern China and transmitted to Hong Kong and 29 other countries with high human morbidity, leading to 8098 confirmed cases and 774 deaths (Hui and Zumla 2019).

SARS-CoV-2 has recently been confirmed in Wuhan City, Hubei Province of China, and spreads to more than 165 countries of the world including India (Li et al. 2020). The virus infection leads to 245,922 confirmed cases and 10,048 deaths worldwide as of March 20, 2020 (ProMED-mail 2020). SARS-CoV-2 and SARS are clinically similar, and recent studies have shown that SARS-CoV-2 is closely related to SARS-CoV (Kumar et al. 2020). The cause and spread of SARS-CoV-2 outbreak are still unclear. Preliminary research has found positive samples for WN-CoV in the wholesale market of Huanan seafood in Wuhan City, but some patients confirmed by the laboratory have not reported visiting this area (Centers for Diseases Control and Prevention (CDC) 2020). Evidence is still emerging; however, information to date shows that transmission from human to human occurs. SARS-CoV-2 infection in patients leads to pneumonia-like symptoms such as fever and difficulty in breathing with radiographs showing invasive pneumonic infiltrates in few cases. The evolution of novel CoVs has been shown to be associated with RNA recombination with the existing CoVs (Andersen et al. 2020). The coronavirus (SARS-CoV-2, MERS, and SARS) infection initially spreads in adults, and the reported symptoms are fever, headache, vomiting, chills, dyspnea, nausea, sore throat, coughing up blood, shortness of breath, myalgia, diarrhea, and malaise (Table 10.1).

Table 10.1 Epidemiology and characteristics of 2019-nCoV, MERS-CoV and SARS-CoV

Characteristics	2019-nCoV	MERS-CoV	SARS-CoV
Genus	*Beta-CoVs, lineage B*	*Beta-CoVs, lineage C*	*Beta-CoVs, lineage B*
Intermediary host	Malayan pangolins	Dromedary camel	Palm civet
Natural reservoir	Bat	Bat	Bat
Origin	Wuhan City, Hubei Province of China	Arabian Peninsula	Guangdong Province, China
Confirmed cases	245,922 (as of March 20, 2020)	2458 (July 31, 2019)	8098
Total deaths	10,048 (as of March 20, 2020)	848	774
Affected countries	168 (as of March 20, 2020)	27	29
Transmission patterns	From animal to human; from human to human	From animal to human; from human to human	From animal to human; from human to human
The predominant receptor	Cellular protease TMPRSS2 and angiotensin-converting enzyme 2	Human dipeptidyl peptidase 4 (DPP4 or CD26)	Human angiotensin-converting enzyme 2 (ACE2)
Cell line susceptibility	Respiratory tract	Respiratory tract, intestinal tract, genitourinary tract, liver, kidney, neurons, monocyte, T lymphocyte and histiocytic cell lines	Respiratory tract; kidney; liver
Viral replication efficiency	Very high	High	Lesser compared to MERS-CoV
Length of nucleotides	29,891	30,119	29,727
Open reading frames (ORFs)	12	11	11
Structural protein	4	4	4
Spike protein (length of amino acids)	1273	1353	1255
Non-structural proteins (NSPs)	16	16	5
Accessory proteins	6	5	8
A characteristic gene order	5′-Replicase ORF1ab, spike (S), envelope (E), membrane (M), and nucleocapsid (N)-3′	5′-Replicase ORF1ab, spike (S), envelope (E), membrane (M), and nucleocapsid (N)-3′	5′-Replicase ORF1ab, spike (S), envelope (E), membrane (M), and nucleocapsid (N)-3′

Severe infection leads to pneumonia, acute respiratory distress syndrome (ARDS), and sometimes multi-organ failure (Paules et al. 2020). Coronavirus infection leads to thrombocytopenia, lymphopenia, and leukopenia with elevated levels of lactate dehydrogenase and liver enzymes (Arabi et al. 2014).

10.2 Treatment of Novel Coronavirus SARS-CoV-2 Infection

According to the World Health Organization (WHO), there is no existing data from randomized clinical trials to advocate any specific anti-nCoV therapy for patients either suspected or diagnosed with SARS-CoV-2. Unlicensed treatments can only be given in the circumstance of ethically approved clinical trials or under the Monitored Emergency Use of Unregistered Interventions System (MEURI), with strict supervision (World Health Organization 2020a). However, researchers have tested a number of FDA-approved drugs against SARS-CoV-2 infection, and these drugs have shown promising antiviral activity in both cell culture and animal models. Some of these drugs are also in clinical trial for SARS-CoV-2 (Li and De Clercq 2020). In the past 2 months, drugs from various classes such as nucleoside analogs, protease inhibitors, and host-targeted agents have been tested to discover an authorized antiviral agent against SARS-CoV-2 infection (Table 10.2). The National Medical Products Administration of China has recently approved fapilavir as the first antiviral medication for the treatment of SARS-CoV-2.

10.2.1 Approved Nucleoside Analogs

At present, two approved nucleoside analogs (ribavirin and favipiravir) demonstrated antiviral activity against SARS-CoV-2 infection (Wang et al. 2020). Nucleoside analogs have efficacy to target RNA-dependent RNA polymerase and inhibit the replication process in a broad range of RNA viruses including beta coronaviruses. Ribavirin was originally licensed for the treatment of HCV (hepatitis C virus) and RSV (respiratory syncytial virus). Ribavirin was also clinically evaluated against MERS and SARS coronaviruses, but the efficacy of the drug is uncertain and associated with severe side effects such as anemia and hypoxia at high doses (Arabi et al. 2019). Similarly, ribavirin was also evaluated against SARS-CoV-2 infection, but the antiviral property of drugs is still not well established against the SARS-CoV-2. In addition, after oral administration, the drug was rapidly absorbed via sodium-dependent nucleoside transporters into the gastrointestinal tract. The drug has oral bioavailability around 64% with large volume of distribution. Drugs such as acetaminophen, acetazolamide, aspirin, acrivastine, and acyclovir are known to decrease the excretion rate that leads to the higher serum level of ribavirin.

Table 10.2 List of drugs that have antiviral activity compounds against SARS-CoV-2

Antiviral agents	Drug targets	Reported mechanism of action
Virus-based treatment approaches		
Favipiravir	RdRp	Inhibits RdRp
Ribavirin	RdRp	Inhibits viral RNA synthesis and mRNA capping
Penciclovir	RdRp	Inhibits RdRp
Remdesivir (GS-5734)	RdRp	Terminates the non-obligate chain
Lopinavir	3CLpro	Inhibits 3CLpro
Ritonavir	3CLpro	Inhibits 3CLpro
Darunavir and cobicistat	3CLpro	Inhibits 3CLpro
ASC09F (HIV protease inhibitor)	3CLpro	Inhibits 3CLpro
Nafamostat	Spike glycoprotein	Inhibits spike-mediated membrane fusion
Griffithsin	Spike glycoprotein	Inhibits spike-mediated membrane fusion
Arbidol (umifenovir)	–	–
Oseltamivir	–	–
Host-based treatment approaches		
Recombinant interferons	Interferon response	Exogenous interferons
Chloroquine	Endosomal acidification	A lysosomotropic base that appears to disrupt intracellular trafficking and viral fusion events
Nitazoxanide	Interferon response	Induces the host innate immune response to produce interferons by the host's fibroblasts and protein kinase R (PKR) activation

Similarly, coadministration of abacavir may increase the hepatotoxicity of ribavirin. Favipiravir (T-705) was initially authorized for the treatment of influenza. In addition, it is reported that favipiravir is also effective against a number of RNA viruses such as Ebola, Nipah, and enterovirus. Similarly, favipiravir demonstrates efficacy against SARS-CoV-2, and currently the drug is in randomized trials with interferon-α and baloxavir marboxil for SARS-CoV-2 infection (Li and De Clercq 2020). The drug is well absorbed after oral administration in the gastrointestinal tract (98% bioavailability) and metabolized in liver, and the metabolites are excreted in urine. Favipiravir is contraindicated in pregnancy due to its teratogenic effect. Favipiravir also decreases the excretion of angiotensin-converting enzyme inhibitors such as captopril.

10.2.2 Experimental Nucleoside Analogs

Unlike the approved nucleoside analogs, two experimental nucleoside analogs remdesivir (GS-5734) and galidesivir (BCX4430) were also investigated against SARS-CoV-2 infection. Remdesivir and Galidesivir are adenine derivatives which demonstrate broad-spectrum antiviral activity in cell cultures and animal models against RNA viruses such as MERS and SARS (Agostini et al. 2018; Sheahan et al. 2020; Warren et al. 2014). Recently, it is shown that the drug inhibited SARS-CoV-2 by integrating into nascent viral RNA chains that lead to premature termination of viral RNA chains (Wang et al. 2020). Remdesivir is presently in clinical trial for Ebola and SARS-CoV-2 infection (Li and De Clercq 2020).

10.2.3 Approved Protease Inhibitors

Drugs such as disulfiram, lopinavir, and ritonavir have antiviral activity against human coronaviruses. Disulfiram was primarily used for the management of alcohol dependence, and it is reported that the drug also inhibits the papain-like protease of MERS and SARS in cell line models (Agostini et al. 2018). In addition, lopinavir and ritonavir are the HIV protease inhibitors that also have efficacy against 3-chymotrypsin-like protease of MERS and SARS (Sheahan et al. 2020; Kim et al. 2016; Chan et al. 2015). However, the mechanism of action of these protease inhibitors is still controversial. At present, lopinavir and ritonavir are in clinical trial in patients infected with SARS-CoV-2, and the results demonstrate that lopinavir–ritonavir treatment fails to reduce mortality or throat viral RNA load in patients with Covid-19 (Cao et al. 2020). In the same way, nafamostat and griffithsin demonstrated inhibitory activity against spike glycoprotein of coronaviruses (Wang et al. 2020; Barton et al. 2014; O'Keefe et al. 2010).

10.2.4 Host-Targeted Strategies

Several immune modulator drugs such as chloroquine, nitazoxanide, and ribavirin in combination of PEGylated interferon alfa-2a and -2b have shown inhibitory action against SARS-CoV-2 infection (Li and De Clercq 2020). A recent study has shown that chloroquine is more effective to prevent SARS-CoV-2 infection in cell culture model compared to other tested drugs and also in an open-label trial for the treatment of SARS-CoV-2 infection in patients (Fig. 10.1) (Wang et al. 2020).

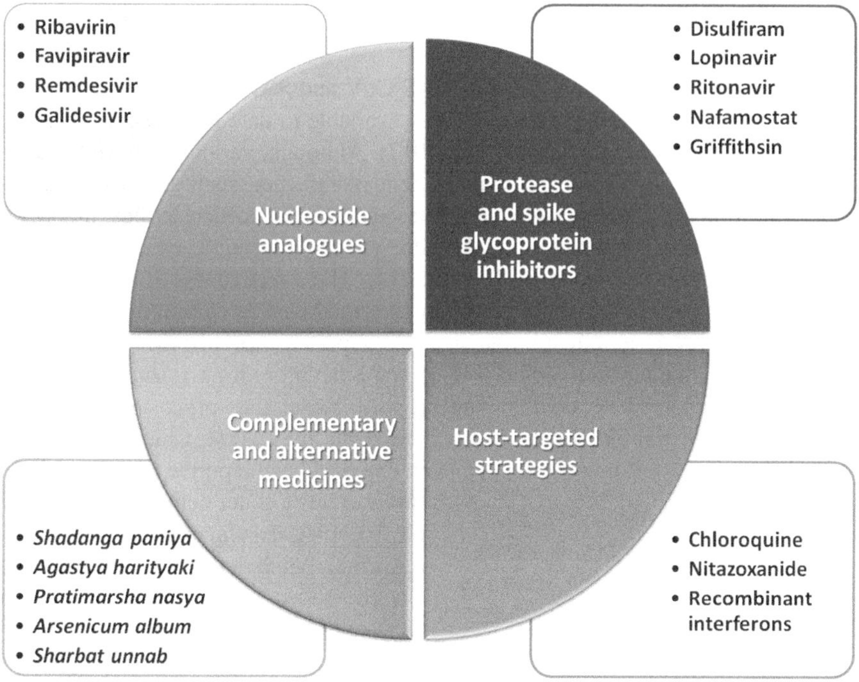

Fig. 10.1 Investigational treatment approaches for COVID-19 infection

10.2.5 Early Supportive Therapy for SARS-CoV-2 Infection

Patients with severe acute respiratory infection should be closely monitored for clinical signs and symptoms like sepsis and rapid respiratory failure to provide immediate supportive treatment in case of severe infection. Supplemental oxygen therapy must be provided in case of shock, hypoxemia, severe acute respiratory infection, and respiratory distress. Similarly, fluid and electrolyte balance should be maintained in infected patients. Antimicrobial agents should be given for the management of pathogen-associated severe acute respiratory infections. The administration of systemic corticosteroids should be avoided in case of viral pneumonia and patients with acute respiratory distress syndrome (World Health Organization 2020b).

10.3 Treatment of MERS-CoV and SARS-CoV Infections

In view of the high mortality rate of MERS-CoV and SARS-CoV infections, there are no licensed antiviral agents or vaccines available to date for successful disease control during outbreaks (Shahani et al. 2017). Although, various therapeutic alternatives have been given for the management of infection such as antiviral drugs (ribavirin, lopinavir, and ritonavir), convalescent plasma, corticosteroids, monoclonal antibodies, intravenous immunoglobulin, and repurposing of existing clinically FDA-approved drugs (Totura and Bavari 2019). The effectiveness of these drugs is not well established and given as symptomatic treatment to control the severity of infection. However, a retrospective analysis reported that ribavirin alone or in combination with interferon-alpha (IFN-ἀ) or lopinavir or ritonavir showed increase in survival compared to patients taking supportive care and corticosteroids against MERS-CoV infection (Falzarano et al. 2013; Chu et al. 2004; Chiou et al. 2005; Loutfy et al. 2003). In the same way, several studies have reported that ribavirin in conjugation with fluoroquinolones and corticosteroids did not demonstrate efficacy in patients with SARS-CoV infection (Table 10.3) (Al-Tawfiq et al. 2014; Omrani et al. 2014; Spanakis et al. 2014; Leong et al. 2004; Muller et al. 2007; Ward et al. 2005; Zhao et al. 2003). Nevertheless, it is concluded that ribavirin alone or in combination with other drugs fails to improve patient outcome against both MERS-CoV and SARS-CoV infections and is associated with adverse effects such as decreased hemoglobin levels, hemolytic anemia, hypoxemia, and metabolic abnormalities. In addition to ribavirin, interferons are also associated with psychiatric disturbances, neutropenia, and systemic adverse effects (Al-Dorzi et al. 2016). Drug repurposing is an eye-catching alternative approach for discovering newer drugs during the virus epidemics for a number of viral infections because various measures are typically skipped during the initial phase of drug design. Recently, researchers have investigated the antiviral potential of various approved drugs as first-line of defense against the newly emerging infectious agents. Repurposing of pharmaceutical agents could be an effective arm to develop potential therapeutics for SARS-CoV-2, MERS-CoV, and SARS-CoV infection (Dyall et al. 2017). The drugs from various classes such as antidiarrheal agents, antimalarial agents, cyclophilin inhibitors, interferons, kinase inhibitors, neurotransmitter inhibitors, anticholinergics, nucleic acid synthesis inhibitors, protease inhibitors, protein synthesis inhibitors, selective estrogen receptor modulators, and sterol metabolism inhibitors have demonstrated in vitro and in vivo antiviral efficacy against MERS-CoV and SARS-CoV infection (Table 10.4) (Dyall et al. 2014, 2017; de Wilde et al. 2011, 2014; Hart et al. 2014; Cinatl et al. 2003; Kindrachuk et al. 2015; Cheng et al. 2015; Saijo et al. 2005; Zhou et al. 2015; Cao et al. 2015).

Table 10.3 Drug regimens used for the treatment of MERS and SARS-CoV infection

Treatment	Outcome
MERS-CoV infection	
Ribavirin (oral/IV) IFN-ἀ 2b Corticosteroids	Late treatment administration. Disease progression delayed and all patients died
Ribavirin (oral/IV) PEGylated IFN-ἀ (IV) ± Corticosteroids	Treatment was given 0–8 days after diagnosis Significant decreases in hemoglobin and absolute neutrophil count (baseline count lower in treatment group) were also reported
Ribavirin (oral/IV) Lopinavir/ritonavir IFN-ἀ-2b	No detectable viral RNA in serum after 2 days of therapy Ribavirin discontinued due to jaundice, hyperbilirubinemia Patients died with septic shock in 2 months
SARS-CoV infection	
Ribavirin (oral/IV) Antibiotics ± Corticosteroids ± Immunoglobulin	No increased positive outcome with ribavirin compared to controls Increased risk of anemia, hypomagnesemia, hypoxia, or bradycardia with ribavirin compared to ribavirin-naive patients
Ribavirin (oral/IV) Lopinavir/ritonavir ± Corticosteroids	Fatality or acute respiratory distress syndrome (ARDS) was reduced significantly from 28.8% to 2.4%
IFN-alfacon-1 ± Corticosteroids ± Antibiotics	Increased oxygen saturation Increased clearance of lung abnormalities Slight increase in creatinine kinase concentrations
Fluoroquinolone (IV) Azithromycin (IV) IFN-α (IM) ± Corticosteroids ± Immunoglobulins ± Thymic peptides/proteins	No increased positive outcome
Quinolone (IV) Azithromycin (IV) ± IFN-α ± Corticosteroids	No increased positive outcome
Levofloxacin Azithromycin ± IFN-α ± Corticosteroids	Increased survival Increased clearance of lung abnormalities

Table 10.4 List of drugs with in vitro and in vivo antiviral activity against MERS-CoV and SARS-CoV

Drugs	Class	Mechanism of action
Loperamide	Antidiarrheal agents	Unknown
Chloroquine, amodiaquine, mefloquine	Antimalarial agents	Chloroquine targets the type II transmembrane serine protease of MERS-CoV For SARS-CoV, chloroquine has also been attributed to a deficit in glycosylation of the receptor angiotensin-converting enzyme 2
Cyclosporine A	Cyclophilin inhibitors	Unknown
IFN-α̇, β1a, γ, IFN-α̇ 2b	Interferons	Inhibits replication of MERS-CoV and SARS-CoV
Imatinib, dasatinib, selumetinib, trametinib, sirolimus	Kinase inhibitors	Imatinib and dasatinib act as entry inhibitor for both MERS and SARS Selumetinib and trametinib block the entry and replication of MERS-CoV via targeting ERK/MAPK signaling pathway. Sirolimus reduced MERS-CoV infection by ~60% via targeting mTOR signaling pathway
Chlorpromazine, triflupromazine, thiethylperazine, promethazine fluphenazine, astemizole, chlorphenoxamine, fluspirilene	Neurotransmitter inhibitors	Chlorpromazine inhibits virus entry, whereas antiviral mechanism of other listed drugs was still not clearly understood
Benztropine	Anticholinergic	Unknown
Ribavirin, mycophenolic acid, mizoribine, gemcitabine	Nucleic acid synthesis inhibitors	Unknown
Camostat mesylate, K11777, E-64-D, lopinavir	Protease inhibitors	Camostat mesylate inhibits TMPSSR2-mediated glycoprotein activation of MERS-CoV and SARS-CoV. K11777 and E-64-D act as attachment inhibitor for both MERS-CoV and SARS-CoV. Lopinavir has been shown to target the main protease (Mpro) of SARS-CoV
Emetine, anisomycin, omacetaxine mepesuccinate	Protein synthesis inhibitors	Unknown
Toremifene citrate, tamoxifen citrate	Selective estrogen receptor modulators	Unknown
Terconazole and triparanol	Sterol metabolism inhibitors	Unknown

10.4 Vaccine for Coronaviruses

Currently, vaccines that can protect against CoV infection are not available. Recently, many groups are involved in vaccine designing using a variety of platforms against CoVs. Some of these approaches have demonstrated effectiveness in animal models. The spike (S) protein present in CoVs acts as a viral antigen and responsible for host–receptor binding and virus internalization and induces robust humoral and cell-mediated responses in humans during infection. The S glycoprotein has been shown to involve in the internalization of other CoVs like SARS by binding to its cellular receptor angiotensin-converting enzyme 2 (ACE2). The S-protein functions in the receptor binding and membrane fusion make it the ideal target for the production of vaccines against CoVs (Wang et al. 2016). Recent experiments have shown that the S-protein vaccine can trigger antibodies to resist virus binding, fusion, and neutralization of SARS-CoV infection. Recently, S-protein-based vaccines such as DNA vaccines, modified vaccinia Ankara (MVA)-based chimeric virus vaccines, subunit vaccines, and virus-like replicon particle (VRP)-based chimeric virus vaccines have been developed, and they demonstrate protective effect against both MERS-CoV and SARS-CoV infections in animal models (Schindewolf and Menachery 2019).

10.5 Complementary and Alternative Medicines (CAM) for Coronaviruses

According to the National Center for Complementary and Integrative Health (NCCIH) of the National Institute of Health (NIH), USA, CAM encompasses various medical methods, such as homeopathy, naturopathy, Ayurveda, medicinal systems, and products originating from traditional medicine. Recently, CAM therapies have shown to be potential therapeutics for the management of virus-associated diseases such as influenza, Japanese encephalitis, hepatitis C, zika, and HIV (Saxena et al. 2017). These medicines also demonstrate efficacy against coronaviruses with minimal reported adverse effects on host cells. Considering the global transmission and fatality rate of SARS-CoV-2 infection, the Government of India, Ministry of Ayush has given recommendations for the use of Indian herbal drugs that practices under the Ayurveda, Homeopathy, and Unani system of medicine for combating coronaviruses (PIB Delhi 2020).

10.5.1 Ayurverdic Medicines for the Treatment of Coronaviruses

Shadanga Paniya is a herbal formulation that mainly comprises *Cyperus rotundus*, *Fumaria indica*, *Vetiveria zizanioides*, *Pterocarpus santalinus*, *Pavonia odorata*, and *Zingiber officinale*). This herbal formulation is recommended for the treatment of symptoms such as high fever, shivering, muscle aches, headache, loss of appetite, dehydration, fatigue, restlessness, excessive thirst, irritability, and burning sensation. In addition, Shadanga Paniya also has antibacterial and antimicrobial activities, and recently, this medicine is recommended by the Ministry of Ayush for the treatment of coronaviruses. Agastya Harityaki is a popular polyherbal Ayurverdic medicine mainly recommended for respiratory problems such as asthma, pneumonia, and chronic bronchitis. The medicine is reported to have antiviral, antibacterial, antifungal, antioxidant, anticarcinogenic, antiaging, antidiabetic, antiulcer, cardioprotective, hepatoprotective, and wound healing properties. Samshamani Vati is used for the treatment of acute to chronic fever and anemia (500 mg twice a day). *Tinospora cordifolia* is the main ingredient and responsible for anti-inflammatory and antipyretic properties of Samshamani Vati. Pratimarsha Nasya (Anu taila/sesame oil) has preventive as well as curative aspect for the treatment of Nasobronchial diseases and enhances the respiratory immunity. The ingredients present in sesame oil are well known for anti-inflammatory, antipyretic, and antibacterial proprieties. Another formulation which comprises Trikatu (Pippali, Marich, and Shunthi) and Tulasi is also recommended by the Ministry of Ayush for the treatment of coronavirus infection in India.

10.5.2 Homeopathic Medicines for the Treatment of Coronaviruses

After the emergence of SARS-CoV-2 in India, the Central Council for Research in Homeopathy (CCRH) has recommended Arsenicum Album 30 as prophylactic medicine against coronavirus infections. It was recommended that one dose of Arsenicum Album 30 should be given daily in empty stomach for 3 days and should be replicated after 1 month on the same schedule in the case of coronavirus infections arising in the population.

10.5.3 Unani Medicines for the Treatment of Coronaviruses

The Government of India has suggested a number of Unani medicines such as Sharbat Unnab (10–20 ml), TiryaqArba (3–5 g), TiryaqNazla (5 g), Habb e IkseerBukhar (2 pills), Sharbat Nazla (10 ml), and Qurs e Suaal (2 tablets) recommended twice a day with lukewarm water. In addition, Arq Ajeeb (4–8 drops) and Khamira Marwareed (3–5 g) are also suggested for the better management of disease. Similarly, medicines like RoghanBaboona/Roghan Mom/Kafoori Balm are also recommended for massage on scalp and chest in case of infection. In addition, the Arq extracted from single Unani drugs like *Azadirachta indica* (Margosa), *Swertiachirata karst* (Indian Gentian), *Trachyspermum ammi sprague* (ajowan), *Cichorium intybus* Linn. (common chicory), *Cyperus scariosus* R. Br. (cypriol), *Borage officinalis* Linn. (Borage), and *Artemisia absinthium* Linn. (common sagewort) along with SharbatKhaksi may be used to combat the infection. In the same way, decoction of Unani drugs such as *Cydonia oblonga* (Quince), *Zizyphus Jujube* Linn. (jujubi), Papaver somniferum (khashkhash), *Cordia myxa* Linn. (Assyrian plum), *Cinnamomum zeylanicum* (cinnamon), *Hyoscyamus niger* (bazrulbanj), *Viola odorata* Linn. (sweet violet), *Borago Officinalis* Linn. (borage), *Myrtus communis* (Barg e Moard), *Lactuca sativa* (Tukhm e kahuMukashar), and *Rosa damascene* (GuleSurkh) are also being suggested for conditions like sore throat during infection.

10.5.4 Herbal Medicines Showing Efficacy Against Coronavirus in Cell Culture Model

It is reported that a number of herbal extract of *Anthemis hyaline*, *Acanthopanacis cortex*, *Citrus sinensis*, *Sophorae radix*, *Sanguisorbae radix*, *Nigella sativa*, and *Torilis fructus* inhibit coronavirus replication in vitro (Ulasli et al. 2014; Kim et al. 2010). Similarly, studies have demonstrated that traditionally used medicinal herbal extracts such as *Cimicifuga rhizoma*, *Coptidis rhizoma*, *Phellodendron cortex*, and *Meliae cortex* are also shown to inhibit CoV replication in cell culture model (Kim et al. 2008). Emodin, a chemical constituent of genus Rheum, and Polygonum are also shown to block the binding of SARS-CoV S-protein to ACE2 in a dose-dependent manner (Ho et al. 2007). In the same manner, *Artemisia annua*, *Lycoris radiate*, *Lindera aggregata*, *lycorine*, and *Pyrrosia lingua* also showed antiviral activity against SARS-CoV (Li et al. 2005). In addition, saikosaponins are natural triterpene glycosides (A, B_2, C, and D) and were tested against HCoV-22E9, and it is found that saikosaponin B_2 inhibits virus attachment and penetration stage to the host cells (Cheng et al. 2006). In the same way, myricetin and scutellarein act as inhibitor of SARS-CoV helicase. Similarly, theaflavin-3,3-digallate (TF3) is a natural polyphenolic compound and demonstrates antiviral activity via targeting 3C-like proteases of SARS-CoV (Chen et al. 2005). Eupatorium fortune is a herbal medicinal plant belonging to Korea, China, and Asian countries with antibacterial and antioxidant activities and anticancer activity. Recent studies have reported that eupatorium fortune can inhibit a number of RNA viruses including human coronaviruses (Choi et al. 2017). As these herbal drugs demonstrated potential antiviral activity against CoVs, there is an urgent need to establish their effectiveness in humans and approve in a dose-dependent manner from various organizations like Food and Drug Administration (FDA). In the absence of specific antiviral agent or vaccine, the use of complementary and alternative drugs may found to be beneficial during humanitarian emergencies.

10.6 Conclusions

The recent outbreak brought SARS-CoV-2 as global concern and emphasizes the significance of restricting infectious agents at international borders. The risk of SARS-CoV-2 outbreaks depends on the characteristics of the virus, including the ability of the virus to transmit among humans, the seriousness of the disease, and the medical or other interventions available to prevent the transmission of virus, such as vaccines or drugs. Although, ribavirin along with corticosteroids and interferons has been tested in patients with SARS and MARS-CoV infections, but the effectiveness and associated severe adverse effects of this regimen is still not yet confirmed. In addition, a large number of drugs from different classes have been used for the management of symptoms associated with the infection and reported to have antiviral activity against both SARS and MARS-CoV infections in animal and cell culture model. However, the antiviral potential of these repurposing drugs are need to be validated in clinical trials in order to developed broad-spectrum therapy for SARS-CoV-2, SARS-CoV, and MARS-CoV infections.

10.7 Future Perspectives

At present, the control of viral spread is critical in order to restrict the transmission of infection. The effective communication among the various organizations such as government, industry, academic, national, and international bodies is crucial in order to prevent the transmission of infection during outbreaks. In the absence of specific therapeutic agent, the study of pathogenesis of infection is imperative to identify newer targets for designing of novel therapeutic agents for future outbreaks. Similarly, the standardization of various complementary and alternative medicines may prove as safe and potential therapeutic strategies for emerging CoVs.

Acknowledgements and Disclosures The authors are grateful to the Vice Chancellor, King George's Medical University (KGMU), Lucknow, India for the encouragement for this work. The authors have no other relevant affiliations or financial involvement with any organization or entity with a financial interest in or financial conflict with the subject matter or materials discussed in the manuscript apart from those disclosed.

Financial support: None.

Conflict of interest: The authors declare no conflict of interest.

References

Agostini ML, Andres EL, Sims AC, Graham RL, Sheahan TP, Lu X, Smith EC, Case JB, Feng JY, Jordan R, Ray AS, Cihlar T, Siegel D, Mackman RL, Clarke MO, Baric RS, Denison MR (2018) Coronavirus susceptibility to the antiviral Remdesivir (GS-5734) is mediated by the viral polymerase and the proofreading Exoribonuclease. MBio 9(2). pii: e00221-18. https://doi.org/10.1128/mBio.00221-18

Al-Dorzi HM, Alsolamy S, Arabi YM (2016) Critically ill patients with Middle East respiratory syndrome coronavirus infection. Crit Care 20:65. https://doi.org/10.1186/s13054-016-1234-4

Al-Tawfiq JA, Momattin H, Dib J, Memish ZA (2014) Ribavirin and interferon therapy in patients infected with the Middle East respiratory syndrome coronavirus: an observational study. Int J Infect Dis 20:42–46. https://doi.org/10.1016/j.ijid.2013.12.003

Andersen KG, Rambaut A, Lipkin WI, Holmes EC, Garry RF (2020) The proximal origin of SARS-CoV-2. Nat Med:1–3. https://doi.org/10.1038/s41591-020-0820-9

Arabi YM, Arifi AA, Balkhy HH, Najm H, Aldawood AS, Ghabashi A, Hawa H, Alothman A, Khaldi A, Al Raiy B (2014) Clinical course and outcomes of critically ill patients with Middle East respiratory syndrome coronavirus infection. Ann Intern Med 160(6):389–397. https://doi.org/10.7326/M13-2486

Arabi YM, Shalhoub S, Mandourah Y, Al-Hameed F, Al-Omari A, Al Qasim E, Jose J, Alraddadi B, Almotairi A, Al Khatib K, Abdulmomen A, Qushmaq I, Sindi AA, Mady A, Solaiman O, Al-Raddadi R, Maghrabi K, Ragab A, Al Mekhlafi GA, Balkhy HH, Al Harthy A, Kharaba A, Gramish JA, Al-Aithan AM, Al-Dawood A, Merson L, Hayden FG, Fowler R (2019. pii: ciz544. [Epub ahead of print]) Ribavirin and Interferon therapy for critically ill patients with Middle East Respiratory Syndrome: a multicenter observational study. Clin Infect Dis. https://doi.org/10.1093/cid/ciz544

Barton C, Kouokam JC, Lasnik AB, Foreman O, Cambon A, Brock G, Montefiori DC, Vojdani F, McCormick AA, O'Keefe BR, Palmer KE (2014) Activity of and effect of subcutaneous treatment with the broad-spectrum antiviral lectin griffithsin in two laboratory rodent models. Antimicrob Agents Chemother 58(1):120–127. https://doi.org/10.1128/AAC.01407-13

Cao J, Forrest JC, Zhang X (2015) A screen of the NIH clinical collection small molecule library identifies potential anti-coronavirus drugs. Antivir Res 114:1–10. https://doi.org/10.1016/j.antiviral.2014.11.010

Cao B, Wang Y, Wen D, Liu W, Wang J, Fan G, Ruan L, Song B, Cai Y, Wei M, Li X, Xia J, Chen N, Xiang J, Yu T, Bai T, Xie X, Zhang L, Li C, Yuan Y, Chen H, Li H, Huang H, Tu S, Gong F, Liu Y, Wei Y, Dong C, Zhou F, Gu X, Xu J, Liu Z, Zhang Y, Li H, Shang L, Wang K, Li K, Zhou X, Dong X, Qu Z, Lu S, Hu X, Ruan S, Luo S, Wu J, Peng L, Cheng F, Pan L, Zou J, Jia C, Wang J, Liu X, Wang S, Wu X, Ge Q, He J, Zhan H, Qiu F, Guo L, Huang C, Jaki T, Hayden FG, Horby PW, Zhang D, Wang C (2020) A trial of Lopinavir-Ritonavir in adults hospitalized with severe Covid-19. N Engl J Med. https://doi.org/10.1056/NEJMoa2001282

Centers for Diseases Control and Prevention (CDC) (2020). https://www.cdc.gov/coronavirus/2019-ncov/about/transmission.html. Accessed 15 Mar 2020

Chan JF, Yao Y, Yeung ML, Deng W, Bao L, Jia L, Li F, Xiao C, Gao H, Yu P, Cai JP, Chu H, Zhou J, Chen H, Qin C, Yuen KY (2015) Treatment with Lopinavir/Ritonavir or Interferon-β1b improves outcome of MERS-CoV infection in a nonhuman primate model of common marmoset. J Infect Dis 212(12):1904–1913. https://doi.org/10.1093/infdis/jiv392

Chen CN, Lin CP, Huang KK, Chen WC, Hsieh HP, Liang PH, Hsu JT (2005) Inhibition of SARS-CoV 3C-like protease activity by theaflavin-3,3′-digallate (TF3). Evid Based Complement Alternat Med 2(2):209–215

Chen N, Zhou M, Dong X, Qu J, Gong F, Han Y, Qiu Y, Wang J, Liu Y, Wei Y, Xia J, Yu T, Zhang X, Zhang L (2020) Epidemiological and clinical characteristics of 99 cases of 2019 novel coronavirus pneumonia in Wuhan, China: a descriptive study. Lancet 395(10223):507–513. . pii: S0140-6736(20)30211-7. https://doi.org/10.1016/S0140-6736(20)30211-7

Cheng PW, Ng LT, Chiang LC, Lin CC (2006) Antiviral effects of saikosaponins on human coronavirus 229E in vitro. Clin Exp Pharmacol Physiol 33(7):612–616

Cheng KW, Cheng SC, Chen WY, Lin MH, Chuang SJ, Cheng IH, Sun CY, Chou CY (2015) Thiopurine analogs and mycophenolic acid synergistically inhibit the papain-like protease of Middle East respiratory syndrome coronavirus. Antivir Res 115:9–16. https://doi.org/10.1016/j.antiviral.2014.12.011

Chiou HE, Liu CL, Buttrey MJ, Kuo HP, Liu HW, Kuo HT, Lu YT (2005) Adverse effects of ribavirin and outcome in severe acute respiratory syndrome: experience in two medical centers. Chest 128(1):263–272

Choi JG, Lee H, Hwang YH, Lee JS, Cho WK, Ma JY (2017) *Eupatorium fortunei* and its components increase antiviral immune responses against RNA viruses. Front Pharmacol 8:511. https://doi.org/10.3389/fphar.2017.00511

Chu CM, Cheng VC, Hung IF, Wong MM, Chan KH, Chan KS, Kao RY, Poon LL, Wong CL, Guan Y, Peiris JS, Yuen KY, HKU/UCH SARS Study Group (2004) Role of lopinavir/ritonavir in the treatment of SARS: initial virological and clinical findings. Thorax 59(3):252–256

Cinatl J, Morgenstern B, Bauer G, Chandra P, Rabenau H, Doerr HW (2003) Treatment of SARS with human interferons. Lancet 362(9380):293–294

Dyall J, Coleman CM, Hart BJ, Venkataraman T, Holbrook MR, Kindrachuk J, Johnson RF, Olinger GG Jr, Jahrling PB, Laidlaw M, Johansen LM, Lear-Rooney CM, Glass PJ, Hensley LE, Frieman MB (2014) Repurposing of clinically developed drugs for treatment of Middle East respiratory syndrome coronavirus infection. Antimicrob Agents Chemother 58(8):4885–4893. https://doi.org/10.1128/AAC.03036-14

Dyall J, Gross R, Kindrachuk J, Johnson RF, Olinger GG Jr, Hensley LE, Frieman MB, Jahrling PB (2017) Middle East Respiratory Syndrome and Severe Acute Respiratory Syndrome: current therapeutic options and potential targets for novel therapies. Drugs 77(18):1935–1966. https://doi.org/10.1007/s40265-017-0830-1

Falzarano D, de Wit E, Rasmussen AL, Feldmann F, Okumura A, Scott DP, Brining D, Bushmaker T, Martellaro C, Baseler L, Benecke AG, Katze MG, Munster VJ, Feldmann H (2013) Treatment with interferon-α2b and ribavirin improves outcome in MERS-CoV-infected rhesus macaques. Nat Med 19(10):1313–1317. https://doi.org/10.1038/nm.3362

Hart BJ, Dyall J, Postnikova E, Zhou H, Kindrachuk J, Johnson RF, Olinger GG Jr, Frieman MB, Holbrook MR, Jahrling PB, Hensley L (2014) Interferon-β and mycophenolic acid are potent inhibitors of Middle East respiratory syndrome coronavirus in cell-based assays. J Gen Virol 95 (Pt 3):571–577. https://doi.org/10.1099/vir.0.061911-0

Ho TY, Wu SL, Chen JC, Li CC, Hsiang CY (2007) Emodin blocks the SARS coronavirus spike protein and angiotensin-converting enzyme 2 interaction. Antivir Res 74(2):92–101

Huang C, Wang Y, Li X, Ren L, Zhao J, Hu Y, Zhang L, Fan G, Xu J, Gu X, Cheng Z, Yu T, Xia J, Wei Y, Wu W, Xie X, Yin W, Li H, Liu M, Xiao Y, Gao H, Guo L, Xie J, Wang G, Jiang R, Gao Z, Jin Q, Wang J, Cao B (2020) Clinical features of patients infected with 2019 novel coronavirus in Wuhan, China. Lancet 395(10223):497–506. pii: S0140-6736(20)30183-5. https://doi.org/10.1016/S0140-6736(20)30183-5

Hui DSC, Zumla A (2019) Severe Acute Respiratory Syndrome: historical, epidemiologic, and clinical features. Infect Dis Clin North Am 33(4):869–889. https://doi.org/10.1016/j.idc.2019.07.001

Kim HY, Shin HS, Park H, Kim YC, Yun YG, Park S, Shin HJ, Kim K (2008) In vitro inhibition of coronavirus replications by the traditionally used medicinal herbal extracts, *Cimicifuga rhizoma*, *Meliae cortex*, *Coptidis rhizoma*, and *Phellodendron cortex*. J Clin Virol 41(2):122–128

Kim HY, Eo EY, Park H, Kim YC, Park S, Shin HJ, Kim K (2010) Medicinal herbal extracts of Sophorae radix, Acanthopanacis cortex, Sanguisorbae radix and Torilis fructus inhibit coronavirus replication in vitro. Antivir Ther 15(5):697–709. https://doi.org/10.3851/IMP1615

Kim UJ, Won EJ, Kee SJ, Jung SI, Jang HC (2016) Combination therapy with lopinavir/ritonavir, ribavirin and interferon-α for Middle East respiratory syndrome. Antivir Ther 21(5):455–459. https://doi.org/10.3851/IMP3002

Kindrachuk J, Ork B, Hart BJ, Mazur S, Holbrook MR, Frieman MB, Traynor D, Johnson RF, Dyall J, Kuhn JH, Olinger GG, Hensley LE, Jahrling PB (2015) Antiviral potential of ERK/MAPK and PI3K/AKT/mTOR signaling modulation for Middle East respiratory syndrome coronavirus infection as identified by temporal kinome analysis. Antimicrob Agents Chemother 59(2):1088–1099. https://doi.org/10.1128/AAC.03659-14

Kumar S, Maurya VK, Prasad AK, Bhatt ML, Saxena SK (2020) Structural, glycosylation and antigenic variation between 2019 novel coronavirus (2019-nCoV) and SARS coronavirus (SARS-CoV). Virus Dis:1–9. https://doi.org/10.1007/s13337-020-00571-5

Leong HN, Ang B, Earnest A, Teoh C, Xu W, Leo YS (2004) Investigational use of ribavirin in the treatment of severe acute respiratory syndrome, Singapore, 2003. Tropical Med Int Health 9 (8):923–927

Li G, De Clercq E (2020) Therapeutic options for the 2019 novel coronavirus (2019-nCoV). Nat Rev Drug Discov. https://doi.org/10.1038/d41573-020-00016-0

Li SY, Chen C, Zhang HQ, Guo HY, Wang H, Wang L, Zhang X, Hua SN, Yu J, Xiao PG, Li RS, Tan X (2005) Identification of natural compounds with antiviral activities against SARS-associated coronavirus. Antivir Res 67(1):18–23

Li Q, Guan X, Wu P, Wang X, Zhou L, Tong Y, Ren R, Leung KSM, Lau EHY, Wong JY, Xing X, Xiang N, Wu Y, Li C, Chen Q, Li D, Liu T, Zhao J, Li M, Tu W, Chen C, Jin L, Yang R, Wang Q, Zhou S, Wang R, Liu H, Luo Y, Liu Y, Shao G, Li H, Tao Z, Yang Y, Deng Z, Liu B, Ma Z, Zhang Y, Shi G, Lam TTY, Wu JTK, Gao GF, Cowling BJ, Yang B, Leung GM, Feng Z (2020) Early transmission dynamics in Wuhan, China, of novel coronavirus-infected pneumonia. N Engl J Med 382:1199–1207. https://doi.org/10.1056/NEJMoa2001316

Loutfy MR, Blatt LM, Siminovitch KA, Ward S, Wolff B, Lho H, Pham DH, Deif H, LaMere EA, Chang M, Kain KC, Farcas GA, Ferguson P, Latchford M, Levy G, Dennis JW, Lai EK, Fish EN (2003) Interferon alfacon-1 plus corticosteroids in severe acute respiratory syndrome: a preliminary study. JAMA 290(24):3222–3228

Muller MP, Dresser L, Raboud J, McGeer A, Rea E, Richardson SE, Mazzulli T, Loeb M, Louie M, Canadian SARS Research Network (2007) Adverse events associated with high-dose ribavirin: evidence from the Toronto outbreak of severe acute respiratory syndrome. Pharmacotherapy 27 (4):494–503

O'Keefe BR, Giomarelli B, Barnard DL, Shenoy SR, Chan PK, McMahon JB, Palmer KE, Barnett BW, Meyerholz DK, Wohlford-Lenane CL, McCray PB Jr (2010) Broad-spectrum in vitro activity and in vivo efficacy of the antiviral protein griffithsin against emerging viruses of the family Coronaviridae. J Virol 84(5):2511–2521. https://doi.org/10.1128/JVI.02322-09

Omrani AS, Saad MM, Baig K, Bahloul A, Abdul-Matin M, Alaidaroos AY, Almakhlafi GA, Albarrak MM, Memish ZA, Albarrak AM (2014) Ribavirin and interferon alfa-2a for severe Middle East respiratory syndrome coronavirus infection: a retrospective cohort study. Lancet Infect Dis 14(11):1090–1095. https://doi.org/10.1016/S1473-3099(14)70920-X

Paules CI, Marston HD, Fauci AS (2020) Coronavirus infections-more than just the common cold. JAMA. https://doi.org/10.1001/jama.2020.0757. [Epub ahead of print]

PIB Delhi (2020) Advisory for corona virus. Homoeopathy for prevention of corona virus Infections. unani medicines useful in symptomatic management of corona virus infection. Posted on: 29 Jan 2020 10:29 AM by PIB Delhi. https://pib.gov.in/PressReleasePage.aspx?PRID=1600895

ProMED-mail (2020) Novel coronavirus (32): China, global, case management, research. https://promedmail.org/. Accessed 20 Mar 2020

Rasmussen SA, Watson AK, Swerdlow DL (2016) Middle East Respiratory Syndrome (MERS). Microbiol Spectr 4(3). https://doi.org/10.1128/microbiolspec.EI10-0020-2016

Saijo M, Morikawa S, Fukushi S, Mizutani T, Hasegawa H, Nagata N, Iwata N, Kurane I (2005) Inhibitory effect of mizoribine and ribavirin on the replication of severe acute respiratory syndrome (SARS)-associated coronavirus. Antivir Res 66(2–3):159–163

Saxena SK, Haikerwal A, Gadugu S, Bhatt ML (2017) Complementary and alternative medicine in alliance with conventional medicine for dengue therapeutics and prevention. Future Virol 12 (8):399–402. https://doi.org/10.2217/fvl-2017-0047

Schindewolf C, Menachery VD (2019) Middle East respiratory syndrome vaccine candidates: cautious optimism. Viruses 11(1). pii: E74. https://doi.org/10.3390/v11010074

Shahani L, Ariza-Heredia EJ, Chemaly RF (2017) Antiviral therapy for respiratory viral infections in immunocompromised patients. Expert Rev Anti-Infect Ther 15(4):401–415. https://doi.org/10.1080/14787210.2017.1279970

Sheahan TP, Sims AC, Leist SR, Schäfer A, Won J, Brown AJ, Montgomery SA, Hogg A, Babusis D, Clarke MO, Spahn JE, Bauer L, Sellers S, Porter D, Feng JY, Cihlar T, Jordan R, Denison MR, Baric RS (2020) Comparative therapeutic efficacy of Remdesivir and combination lopinavir, ritonavir, and interferon beta against MERS-CoV. Nat Commun 11(1):222. https://doi.org/10.1038/s41467-019-13940-6

Shehata MM, Gomaa MR, Ali MA, Kayali G (2016) Middle East respiratory syndrome coronavirus: a comprehensive review. Front Med 10(2):120–136. https://doi.org/10.1007/s11684-016-0430-6

Skariyachan S, Challapilli SB, Packirisamy S, Kumargowda ST, Sridhar VS (2019) Recent aspects on the pathogenesis mechanism, animal models and novel therapeutic interventions for Middle East respiratory syndrome coronavirus infections. Front Microbiol 10:569. https://doi.org/10.3389/fmicb.2019.00569

Song Z, Xu Y, Bao L, Zhang L, Yu P, Qu Y, Zhu H, Zhao W, Han Y, Qin C (2019) From SARS to MERS, thrusting coronaviruses into the spotlight. Viruses 11(1). pii: E59. https://doi.org/10.3390/v11010059

Spanakis N, Tsiodras S, Haagmans BL, Raj VS, Pontikis K, Koutsoukou A, Koulouris NG, Osterhaus AD, Koopmans MP, Tsakris A (2014) Virological and serological analysis of a recent Middle East respiratory syndrome coronavirus infection case on a triple combination antiviral regimen. Int J Antimicrob Agents 44(6):528–532. https://doi.org/10.1016/j.ijantimicag.2014.07.026

Totura AL, Bavari S (2019) Broad-spectrum coronavirus antiviral drug discovery. Expert Opin Drug Discov 14(4):397–412. https://doi.org/10.1080/17460441.2019.1581171

Ulasli M, Gurses SA, Bayraktar R, Yumrutas O, Oztuzcu S, Igci M, Igci YZ, Cakmak EA, Arslan A (2014) The effects of Nigella sativa (Ns), Anthemis hyalina (Ah) and Citrus sinensis (Cs) extracts on the replication of coronavirus and the expression of TRP genes family. Mol Biol Rep 41(3):1703–1711. https://doi.org/10.1007/s11033-014-3019-7

Wang Q, Wong G, Lu G, Yan J, Gao GF (2016) MERS-CoV spike protein: targets for vaccines and therapeutics. Antivir Res 133:165–177. https://doi.org/10.1016/j.antiviral.2016.07.015

Wang M, Cao R, Zhang L, Yang X, Liu J, Xu M, Shi Z, Hu Z, Zhong W, Xiao G (2020) Remdesivir and chloroquine effectively inhibit the recently emerged novel coronavirus (2019-nCoV) in vitro. Cell Res. https://doi.org/10.1038/s41422-020-0282-0. [Epub ahead of print]

Ward SE, Loutfy MR, Blatt LM, Siminovitch KA, Chen J, Hinek A, Wolff B, Pham DH, Deif H, LaMere EA, Kain KC, Farcas GA, Ferguson P, Latchford M, Levy G, Fung L, Dennis JW, Lai EK, Fish EN (2005) Dynamic changes in clinical features and cytokine/chemokine responses in SARS patients treated with interferon alfacon-1 plus corticosteroids. Antivir Ther 10 (2):263–275

Warren TK, Wells J, Panchal RG, Stuthman KS, Garza NL, Van Tongeren SA, Dong L, Retterer CJ, Eaton BP, Pegoraro G, Honnold S, Bantia S, Kotian P, Chen X, Taubenheim BR, Welch LS, Minning DM, Babu YS, Sheridan WP, Bavari S (2014) Protection against filovirus diseases by a novel broad-spectrum nucleoside analogue BCX4430. Nature 508(7496):402–405. https://doi.org/10.1038/nature13027

de Wilde AH, Zevenhoven-Dobbe JC, van der Meer Y, Thiel V, Narayanan K, Makino S, Snijder EJ, van Hemert MJ (2011) Cyclosporin A inhibits the replication of diverse coronaviruses. J Gen Virol 92(Pt 11):2542–2548. https://doi.org/10.1099/vir.0.034983-0

de Wilde AH, Jochmans D, Posthuma CC, Zevenhoven-Dobbe JC, van Nieuwkoop S, Bestebroer TM, van den Hoogen BG, Neyts J, Snijder EJ (2014) Screening of an FDA-approved compound library identifies four small-molecule inhibitors of Middle East respiratory syndrome coronavirus replication in cell culture. Antimicrob Agents Chemother 58(8):4875–4884. https://doi.org/10.1128/AAC.03011-14

World Health Organization (2020a) Novel Coronavirus (2019-nCoV) situation report-14. https://www.who.int/docs/default-source/coronaviruse/situationreports/20200203-sitrep-14-ncov.pdf. Accessed 3 Feb 2020

World Health Organization (2020b) Clinical management of severe acute respiratory infection when novel coronavirus (nCoV) infection is suspected. Interim guidance. 12 Jan 2020

Zhao Z, Zhang F, Xu M, Huang K, Zhong W, Cai W, Yin Z, Huang S, Deng Z, Wei M, Xiong J, Hawkey PM (2003) Description and clinical treatment of an early outbreak of severe acute respiratory syndrome (SARS) in Guangzhou, PR China. J Med Microbiol 52(Pt 8):715–720

Zheng J, Hassan S, Alagaili AN, Alshukairi AN, Amor NMS, Mukhtar N, Nazeer IM, Tahir Z, Akhter N, Perlman S, Yaqub T (2019) Middle East Respiratory syndrome coronavirus seropositivity in Camel handlers and their families, Pakistan. Emerg Infect Dis 25(12). https://doi.org/10.3201/eid2512.191169

Zhou Y, Vedantham P, Lu K, Agudelo J, Carrion R Jr, Nunneley JW, Barnard D, Pöhlmann S, McKerrow JH, Renslo AR, Simmons G (2015) Protease inhibitors targeting coronavirus and filovirus entry. Antivir Res 116:76–84. https://doi.org/10.1016/j.antiviral.2015.01.011

Chapter 11
Prevention and Control Strategies for SARS-CoV-2 Infection

Nishant Srivastava and Shailendra K. Saxena

Abstract The population of 168 countries all over the world is struggling with the outbreak of COVID-19. The outbreak is declared as pandemic and public health emergency of international concern declared by WHO. SARS-CoV-2 responsible for the present health emergency exhibited close resemblance with SARS-CoV. Both the viruses are zoonotic and belong to a large family of viruses *Coronaviridae*. The complete virus particle is made up of four major structural proteins, namely spikes (S), nucleocapsid (N), membrane (M), and envelope (E) encoded by virus genome. The S protein of virus shows similarity to S protein of SARS-CoV. COVID-19 spreads from person to person, and this makes it more vulnerable for causing infection. Several efforts are taken to find prevention strategies for COVID-19. Researchers across the globe are working to find effective vaccination for SARS-CoV-2. There is no vaccine or medication available till date for COVID-19. Preventive measures such as social distancing, awareness, maintenance of hygiene, isolation, and movement restrictions can help in control of COVID-19 spread. Proper sanitization and cleaned and sanitized public transport can be effective in inhibiting the spread of the virus. In the present situation of medical emergency, cooperation and support by following advices from the WHO and government only facilitate everyone to come over.

Keywords Coronavirus · COVID-19 · Vaccine · SARS-CoV-2

Nishant Srivastava and Shailendra K. Saxena contributed equally as first author.

N. Srivastava (✉)
Department of Biotechnology, Meerut Institute of Engineering and Technology, Meerut, India

S. K. Saxena
Centre for Advanced Research (CFAR), Faculty of Medicine, King George's Medical University (KGMU), Lucknow, India
e-mail: shailen@kgmcindia.edu

S. K. Saxena (ed.), *Coronavirus Disease 2019 (COVID-19)*, Medical Virology: from Pathogenesis to Disease Control, https://doi.org/10.1007/978-981-15-4814-7_11

Abbreviations

BW CoV	Beluga Whale CoV
CoV	Coronavirus
COVID	Coronavirus disease
HCoV	Human coronavirus
IBV	Infectious bronchitis virus
MERS	Middle East respiratory syndrome
PEDV	Porcine epidemic diarrhoea virus
SADS-CoV	Swine acute diarrhoea syndrome-CoV
SARS	Severe acute respiratory syndrome
TGEV	Transmissible gastroenteritis virus
WHO	World Health Organization

11.1 Introduction

The SARS-COV 2 or COVID-19 is emerged as global pandemic declared by the WHO with 184,976 reported cases across 159 countries until March 18, 2020 and accounts for 7529 deaths globally (WHO). The severity of COVID-19 can be easily understood by the exponentially increasing cases worldwide. The virus affects respiratory system like other influenza viruses and appears as a major threat throughout the world after 1918 Spanish flu (H1N1) outbreak. COVID-19 is one of the highly infectious diseases with the ability to affect a large population globally and can cause severe impact on socioeconomic stability of the world. The emergence of SARS-CoV traces back to year 2003 from China, and again another mutant emerged in 2012 known as MERS from Saudi Arabia. All the three highly infectious strains of CoV are found to be zoonotic and transmitted from animals to people.

The SARS-CoV-2 is also believed to be spread from fish market of Wuhan China in November 2019 and was first reported in the last week of December 2019 in 59 people at Wuhan, Hubei, China. The strain was completely novel and unknown to scientific fraternity at the time of its outbreak. The novel coronavirus or COVID-19 is highly infectious and spread so fast through people-to-people contact. The emergence and re-emergence of zoonotic viral strains pose immense threat to the human population and need to addressed strategically with rapid response (Menachery et al. 2018). The SARS-CoV-2 is a novel coronavirus and is classified into virus family of *Coronaviridae*. Coronaviruses belong to large *Coronaviridae* family of viruses causing infection leads from common cold to severe illness and respiratory diseases. As per the studies conducted by various research groups worldwide, the COVID-19 genome is in close similarity with bat coronavirus, and it belongs to beta coronavirus group of *Coronaviridae* family. The brief classification of the SARS-CoV-2 is depicted in Fig. 11.1 (Tekes and Thiel 2016; Ashour et al. 2020), and some of the major human illnesses causing CoVs are given in Table 11.1.

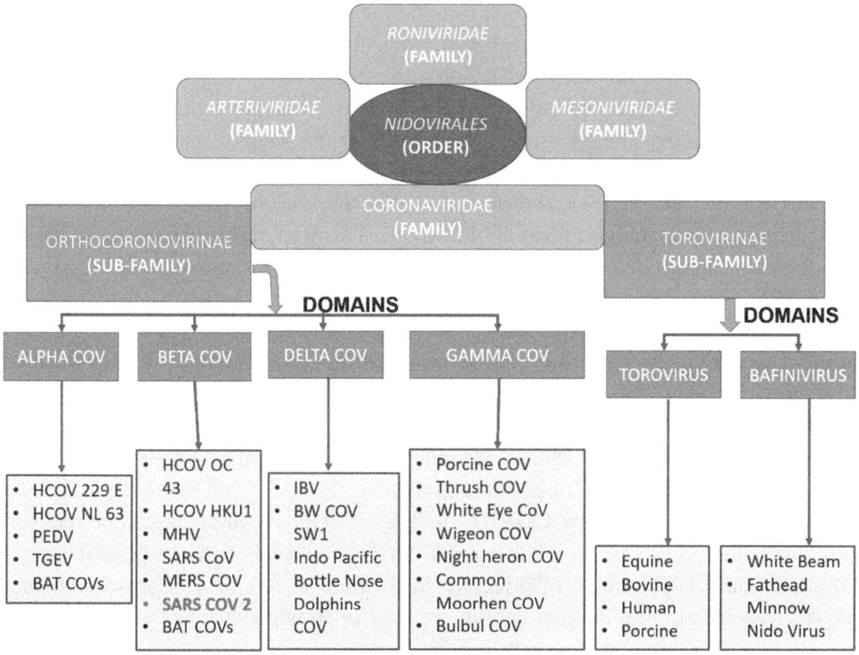

Fig. 11.1 Classification of SARS-CoV-2

Table 11.1 Coronaviruses causing infection in humans

S. No.	Coronavirus	Strain	Disease
1	229 E	Alpha CoV	Common cold and flu
2	NL 63		
3	OC 43	Beta CoV	
4	HKU1		
5	MERS-CoV		Flu, severe respiratory syndrome, pneumonia
6	SARS-CoV		Fever, cough, severe acute respiratory syndrome, bronchitis
7	SARS-CoV-2/ COVID-19		Severe acute respiratory syndrome, fever, dry cough, kidney failure, high transmission between person to person
8	IBV	Delta CoV	Respiratory disease

The emergence of COVID-19 has drawn very much attention of researchers and health professionals because of its high infection potential and novel structure. Researchers found the stability of SARS-CoV-2 on various surfaces and compared the same with SARS-CoV-1. The experimental data under varied circumstances on different surfaces are found to be similar for both the strains (van Doremalen et al. 2020). The experimental observation shows that alterations in the epidemiologic characteristics of both the viruses may arise due to other reasons. The high viral concentration in the upper respiratory tract in the patient and the potential of

COVID-19-infected individual to carry and spread the virus while being in asymptomatic condition are some probable reasons for high infection rate, making its control challenging (Bai et al. 2020; Zou et al. 2020).

COVID-19 is a single-stranded (ss) RNA virus consisting of 26–36 kb positive sense RNA genome. The complete virus particle is made up of four major structural proteins, namely spikes (S), nucleocapsid (N), membrane (M), and envelope (E) encoded by virus genome (Forni et al. 2017). The size of SARS-CoV-2-encoded proteins shows similarity to bat SARS-CoV. The important difference is longer length and structure of S protein of SARS-CoV-2 in comparison to SARS-CoV and bat SARS-CoV observed by researchers. These S proteins are very crucial for receptor binding, membrane fusion, internalization of the SARS-CoV-2, tissue tropism, and host array. This S protein may be utilized as vital target for vaccine development (Menachery et al. 2016; Ji et al. 2020; Kumar et al. 2020).

Extensive efforts are already taken to control COVID-19 spread and for the development of effective vaccines around the world. Scientists are working round the clock individually and in collaboration to get some effective solution for the severe pandemic occurred by COVID-19. The COVID-19 outbreak can be reduced and controlled only by maintaining self-hygiene, social distancing, and strong immunity until any vaccine or effective medication is found. The present chapter provides an insight into the prevention and control strategies for COVID-19 including vaccine development and control measures.

11.2 SARS-CoV-2 Vaccine Development

COVID-19 contains the largest RNA genome and has spike proteins which play very important role in host–virus interaction and infection. After entering into host cells, the viral genome translates into two large precursors, poly-proteins known as PP1a and PP1ab. These precursors further processed by ORF1a-encoded viral proteinases, PL pro (papain-like proteinases) and 3CL pro (3C-like proteinases) into 16 mature non-structural proteins, namely nsp1 to nsp16. These nsps perform several important functions in viral RNA replication and transcription, RNA polymerase, RNA-processing enzymes such as poly (U)-specific endoribonuclease, protease, helicase, 3′–5′ exoribonuclease, ribose 2′-O methyltransferase, adenosine diphosphate-ribose-1″-phosphatase, and cyclic nucleotide phosphodiesterase (Narayanan et al. 2015). Additionally, these nsps have a major role in viral RNA replication and transcription. Due to the absence of proofreading mechanism during RNA recombination process in pre-existed coronaviruses strains, it may account to be responsible for the evolution of SARS-CoV-2. The S gene-encoding spike glycoproteins have the maximum recombination frequency (Ji et al. 2020; Kumar et al. 2020). Out of the three major proteins forming viral envelope, S and M are glycol proteins and the E is non-glycosylated protein. M and E proteins are essential for virus assembly, morphogenesis, and budding. The M protein comprises short N terminal glycosylated ectodomain with a long C terminal domain and three membrane domains. On the other hand, S glycoprotein is a type 1 fusion viral protein which includes two heptad repeat regions HR-N and HR-C and forms protein ectodomain-surrounded coiled structure. Additionally, the S protein cleaves into two subunits S1 and S2 and facilitates receptor binding (at domain 270–510) and membrane fusion, respectively (Tripet et al. 2004; Yuen et al. 2007; Fehr and Perlman 2015; Narayanan et al. 2015; Kumar et al. 2020; Wan et al. 2005). Considering the importance of S protein in attachment with host cell and variation in sequence of S protein from SARS-CoV-1, S protein may be considered potential candidate for vaccine development. Systematic assessment identified that 380 amino acid substitutions between SARS-CoV-1, SARS like bat CoV, and SARS-CoV-2 may be responsible for functional and pathogenic divergence of SARS-CoV-2 (Wu et al. 2020).

The COVID-19 virus is traced till death in non-survivors, and the longest virus shedding of 37 days has been observed in survivors (Zhou et al. 2020). It shows the severity of COVID-19 and its high infection rate especially in elderly and immunocompromised or weak immunity individuals. Several pharmaceutical R&D units and researchers are working to develop vaccine for SARS-CoV-2 pandemic. Researchers are putting all efforts to find a solution for combating this novel coronavirus. Researchers across the globe are trying to develop SARS-CoV-2 vaccine by using approaches like whole virus vaccine, antibody vaccine, DNA vaccine, recombinant protein subunit vaccine, and mRNA vaccine (Dresden 2020). Currently, there is no approved vaccine available for SARS-CoV-2. Recently, mRNA 1273 investigational vaccine has been developed by NIAID scientists in collaboration with the biotechnology company Moderna, Inc., based in Cambridge, MA. The clinical trial began at Kaiser Permanente Washington Health Research Institute (KPWHRI) in Seattle, and part of National Institute of Allergy and Infectious Diseases (NIAID), National Institutes of Health, is funding the trial on mRNA platform approved for human trial after its promising response in animal models (NIH). A team of researchers from The Hong Kong University of Science and Technology, Hong Kong, reported high genetic similarity between SARS-CoV (outbreak in 2003) and SARS-CoV-2. This genetic similarity leads researchers to determine experimental data for SARS-CoV-1 B-cell and T-cell epitopes derived from S and N proteins which map identically to SARS-CoV-2. In comparison to the non-structural proteins, the T-cell response against the structural proteins has been reported to be the most immunogenic in peripheral blood mononuclear cells of recovering SARS-CoV patients. Additionally, T-cell responses against the S and N proteins have been reported to be the most dominant and long-lasting (Li et al. 2008; Channappanavar et al. 2014; Ahmed et al. 2020). As no mutation was observed in the SARS-CoV-2 proteins and long-lasting T-cell response against S protein, immune targeting of these epitopes may provide protection against COVID-19 or SARS-CoV-2 (Ahmed et al. 2020). Similar study was also reported by researchers from CFAR, King George's Medical University, Lucknow, India, in which researchers found glycosylated S protein site as potential target for vaccine development due to the genetic similarity with SARS-CoV and Bat CoV (Kumar et al. 2020). Table 11.2

Table 11.2 Vaccine under development to combat SARS-CoV-2

S. No.	Platform	Type of vaccine	Developer	Current stage (clinical/ regulatory)
1	DNA	DNA plasmid	Zydus Cadilla	Preclinical
2	DNA	INO4800-DNA plasmid	Inovio Pharmaceutical	Preclinical
3	RNA	mRNA 1273	Moderna Inc. and NIAID	Clinical trial phase 1
4	RNA	mRNA	Curevac	Preclinical
5	Live attenu-ated virus	Deoptimized live attenuated virus	Codagenix	Preclinical
6	RNA	mRNA BNT162	Pfizer and BioNTech	Preclinical
7	Recombinant protein subunit	S protein	University of Georgia	Preclinical
8	Recombinant protein subunit	S protein	Novavax	Preclinical
9	Recombinant protein subunit	S protein	University of Queensland	Preclinical
10	Recombinant protein subunit	S protein	Clover Biopharmaceuticals	Preclinical
11	Whole virus	Live attenuated virus	Johnson & Johnson	Preclinical
12	Whole virus	Live attenuated virus	Codagenix	Preclinical
13	Protein subunit	Full-length S trimers/ nanoparticle + matrix M	Novavax	Preclinical
14	Protein subunit	S protein (baculovirus production)	Sanofi Pasteur	Preclinical
15	Replicating viral vector	Measles vector	Zydus Cadila	Preclinical
16	Replicating viral vector	Horsepox vector	Tonix Pharma/Southern Research	Preclinical
17	Protein subunit	S protein clamp	GSK Pharma	Preclinical
18	Protein subunit	S1 protein	Baylor, New York Blood Center, Fudan University	Preclinical
19	Antibody	Antibody-based vaccine	Eli Lilly with AbCelerra	Screening

Source: WHO, Medical News Today, precisionvaccination.com, NIH, Pfizer.com

comprise of some of the vaccines under development in various pharma and R&D laboratories with the stage of vaccine.

The above given set of data is few of ongoing search for potential vaccine. Vaccine development is one of the typical processes and requires lot of efforts and time. There are several vaccines under development for SARS-CoV-1 and MERS-CoV, and researches are still going on at various stages of clinical trials. The available advanced technology for virus genome sequencing and global emergency of pandemic expedite the process to develop vaccine at earliest. The potential vaccine may take more than a year to come on the market after getting all approvals. As once the researchers develop a vaccine, it must get approved from various agencies like FDA after which the vaccine is sent to three phases of clinical trial. In phase 1, a small group of people are selected to evaluate the safety and immune response for vaccine. After successful trial in phase 1, the vaccine is tested for phase 2 clinical trial on approximately few hundred people to analyse the dosage. In phase 3, the effectiveness and safety evaluation of the vaccine will be evaluated on a large population. Once the vaccine passes all the three phases, it successfully gets approval from the FDA and is released for use to combat against the pathogen.

11.3 Types of Coronavirus Vaccines

Since the inception of deadly coronavirus such as SARS-CoV in 2003–2004, the human coronaviruses drew very much attention of the researchers worldwide to find a solution. The emergence and transfer of zoonotic pathogens in humans become a great threat for human population. After a decade of SARS outbreak, MERS is emerged as another severe threat and continued infecting after its discovery in 2012 (Stockman et al. 2006). In 2020 coronavirus emerged in a new form and is standing in front of the whole humanity as a severe threat in the form of pandemic accountable for many lives across the globe. The timely containment and slow rate of transmission of SARS-CoV-1 and MERS helped to control its larger impact (Menachery et al. 2018). In context of SARS and MERS outbreaks, many vaccines were developed and tested. The majority of coronavirus vaccines studied or under development so far broadly comprise the following types:

1. Whole virus vaccine/live attenuated vaccine
2. Antibody-based vaccine
3. Small subunit-based vaccine
4. Vector-based vaccine
5. Nucleic acid-based vaccine

Live attenuated vaccines contain weakened or dead virus to induce immune response in individual recipient. The live virus vaccine sometimes develops complications in the receiver, but it provides a long-term immune response for the specific virus. Antibody-based vaccines contain monoclonal antibodies which are strain specific and provide limited range of protection. As per the studies carried out on coronavirus, it mutated many times and in the last 15–18 years, some species emerged as deadly virus. The protein subunit-based vaccines are one of the safest vaccines, providing protection to a wide range of virus strains by targeting viral S and N proteins which facilitate attachment and interaction of virus to host cells.

Table 11.3 Coronavirus vaccines

S. No.	Platform	Type of vaccine	Target virus	Developer	Current stage	References
1	DNA	INO-4700 DNA plasmid vaccine expresses S protein	MERS-CoV	Inovio Pharmaceuticals	Clinical trial phase 2	Inovio.com
2	DNA	GLS-5300	MERS-CoV	GeneOne Life Science and Inovio	Clinical trial phase 1	Shen et al. (2019)
3	Vector (adeno virus)	ChAdOx1 S protein antigen	MERS CoV	University of Oxford	Clinical trial phase 1	Alharbi et al. (2017)
4	Vector	Ad5-SARS-CoV vectors	SARS-CoV	University of Pittsburgh	Animal trial	Gao et al. (2003)
5	Live attenuated	UV-inactivated purified SARS-CoV	SARS-CoV	National Institute of Infectious Diseases Japan	Subclinical	Takasuka et al. (2004)

Nucleic acid-based vaccines also known as DNA or RNA vaccine are also safe for the recipients and provide long-term immunity. The mRNA-based vaccine for COVID-19 already sent for phase 1 clinical trial. Some of the coronavirus vaccines are listed in Table 11.3.

The research reported by various individual groups shows that CoV S subunit protein provides complete protection than SARS-CoV live attenuated vaccines, full-length S protein, and DNA-based S protein vaccines (Buchholz et al. 2004). The S protein gene of coronavirus is a highly preferable target of antigenicity as exhibited by the reactivity with convalescent SARS patient sera. It shows precise binding to soluble ACE-2 receptor. Additionally, it promotes antibody-dependent viral access in Raji B cells of human refractory B Cells and provokes defense against a impute infection in an animal model (Liu et al. 2020).

11.4 Efficacy and Effectiveness of Coronavirus Vaccines

The efficacy and effectiveness of vaccine are the most important criteria for the development of vaccine. The efficacy of a vaccine is defined as "The reduction percentage of disease in vaccinated or recipients' group in comparison to unvaccinated groups under optimal condition." Whereas, the effectiveness of a vaccine is defined as "The percentage control achieved for particular disease in vaccinated group compared to unvaccinated group at real world in natural environment." The efficacy of vaccine is studied when the vaccine is in clinical trial, whereas the effectiveness is studied after the release of vaccine commercially or to mass after all approvals.

The efficacy of different types of coronavirus vaccines is studied for their potential application in the prevention of the disease. Various types of coronavirus vaccines are under research like inactivated coronavirus, live attenuated coronavirus, S-protein-based, vectored vaccines, DNA vaccines, and combination vaccines. Vaccines for combating several animal-based CoVs have been developed and demonstrated to be efficient in averting viral contagion. Vaccines for human CoVs, especially deadly SARS-CoV-1, MERS-CoV, and the present pandemic COVID-19, are under development in various levels of preclinical, subclinical, and clinical trials. The study of virus-like particle vaccine in mice and whole virus vaccine in non-human primates and ferrets exhibits immune response against CoV infection, but later on vaccinated animals unveiled immunopathological lung disease (Tseng et al. 2012). Another study in canine CoV vaccine exhibited that the vaccination with inactivated CoV can expressively reduce the viral replication followed with the reduction in occurrence of clinical symptoms and disease due to virulent CoV infection (Fulker et al. 1995). In a study carried out on rhesus monkey with SARS-CoV, inactivated vaccine exhibited effective concomitant humoral and mucosal immunity against SARS-CoV infection. The rhesus monkey was inoculated with a varied dose (0.5, 5, 50, and 5000 μg) of vaccine and provided with booster dose after a week. Afterwards the animals were exposed to NS-1 strain of SARS-CoV, the vaccinated monkeys had not shown any systematic side effects neither any clinical symptoms. The vaccine in study shows promising results on monkeys and may be further tested under clinical trials on humans (Zhou et al. 2005).

Various therapeutic efforts were adopted for the CoV vaccine globally; however, no solution in form of promising treatment or vaccination is identified till date. Progressive efforts on developed animal models and under clinical trials are ongoing to find an effective vaccine that is underway. As discussed in the above sections, targeting S glycoproteins of the virus and its subunits responsible for host interaction may be effective target for CoV vaccine for humans. As per research conducted on various CoV genetic structure, the S protein genetic make-up is found to be similar in SARS-CoV, bat SARS-CoV, and SARS-CoV-2. Previously conducted studies on SARS-CoV and MERS-CoV vaccine development based on S glycoproteins of virus can provide insight into effective and efficient vaccine development against deadly strains of corona.

11.5 Control Strategies of Coronavirus Infection

The viral diseases are emerging as serious threat to public health in the past 20 years. Several viral epidemics identified as potential health hazard such as SARS-CoV, H1N1, H5N1 influenza, MERS-CoV, Zika, Ebola, and now COVID-19 erupt as pandemic affecting approximately 168 countries around the globe. The outbreak of SARS-CoV-2 is identified as public health emergency of international concern declared by WHO. The reported potential of human-to-human transmission of the virus makes it a more vulnerable threat for a large population. COVID-19 is a serious global health issue, and scientists across the world are working tirelessly to find its prevention and therapeutic strategies. As of now, no medication or vaccine is available for COVID-19. The symptomatic treatment in mild infections and oxygen therapy in critical cases are found to be effective. There are some reports providing positive response from the use of individual and combination of drugs like ritonavir, chloroquine, lopinavir, BCX-4430 (salt form of galidesivir), nitazoxanide, and ribavirin (Liu et al. 2020). But no fundamental or approved evidence is available for the use of these drugs till date from any international organization like the WHO or FDA.

Presently, only precautionary measures and efficient health response from governments, doctors, and public can only prevent COVID-19 infection from spreading. Person-to-person transmission of virus is critical, and super spreading events can occur in public gathering. Some of the important steps which need to be taken to prevent spreading of COVID-19 in population are as follows:

(a) Isolation of the affected person and individual travelling from affected countries or potential carriers. As some of the studies suggested that COVID-19 can be spread via asymptomatic carriers too, and it is a dangerous situation.
(b) Imposing travel restrictions from and to the affected countries.
(c) Blocking transmission by maintaining high-level hygienic condition in home and surroundings.
(d) Avoid social gathering, as it inhibits its geometrical progression and flatten the curve.
(e) Spreading awareness among public.
(f) Use of masks and protective clothes by infected, elderly, and immunocompromised individuals to avoid spread of infection or from protecting self from COVID-19.
(g) Maintenance of good immunity, consumption of nutrient diet, consumption of vitamins especially C and E with yoga and exercises help in fighting with COVID-19 infection.
(h) Social distancing, as the virus can transmit from person to person, i.e., maintenance of social distance, is highly recommended.

SARS-CoV-2 is a novel virus, and very less is known about the virus. Precaution is better than cure perfectly fits for COVID-19.

"Natural calamities bring people closer, viral calamities keep people away."

Executive Summary

- As per WHO, 209,839 people are affected by COVID-19 until March 20, 2020.
  - COVID-19 is accountable for 8778 death and is present in 168 countries worldwide.

(continued)

- Several studies are going on for exploring efficient medication and vaccine for COVID-19.
  - S-glycoprotein-based and mRNA-based vaccines show prominent response.
- SARS-CoV-2 shows similarity to SARS-CoV and bat-CoV.
  - The similarity in genome structure especially in S protein can be utilized for vaccine development and alternate medications which are effective on SARS-CoV.
- Precaution is only cure is true in the case of COVID-19.
  - Maintenance of social distancing, hygiene, and isolation.
  - Social awareness by government, media, NGOs, and responsible citizens.
  - Isolation of infected, suspected, and possible asymptomatic carriers.
  - Total lock down in highly affected places.
  - Proper maintenance of clean and sanitized environment.
- COVID-19 is a novel virus, and proper precautions is the only cure.
  - Support to government and responsible agencies by following advisory and guidelines.
  - Follow all the instructions and guidelines released by the WHO.

11.6 Conclusions

COVID-19 or SARS-CoV-2 is a novel virus having close genetic resemblance with SARS-CoV and bat-CoV. The virus is highly infectious and can transfer from person to person. As predicted previously by various researchers worldwide, COVID-19 emerged as pandemic. The present situation clearly calls for the fast and efficient availability of COVID-19 vaccines and therapeutics. Worldwide scientists already are working tirelessly for efficient and effective solution of COVID-19 pandemic; meanwhile, it is moral and social obligation of each individual on the planet to follow the guidelines and advisory from the government, the WHO, and other responsible agencies. Extended support and cooperation from public can only help to control the spread of COVID-19 further.

11.7 Future Perspective

1. An in-depth study to understand SARS-CoV-2 needs to be done.
2. Development of efficient vaccines and therapeutics/antiviral drugs needs to be carried out at the earliest.
3. Fast, cost-effective, and accurate diagnostic kits need to be developed.
4. The virus exhibits strong opportunity to study viral mutation pattern and their fast emergence as novel viral particle.
5. Nanomaterial-based drug delivery and diagnostics need to be explored for efficient and effective delivery and diagnostics.

References

Ahmed SF, Quadeer AA, McKay MR (2020) Preliminary identification of potential vaccine targets for the COVID-19 coronavirus (SARS-CoV-2) based on SARS-CoV immunological studies. Viruses 12(3):254

Alharbi NK, Padron-Regalado E, Thompson CP, Kupke A, Wells D, Sloan MA, Grehan K, Temperton N, Lambe T, Warimwe G, Becker S, Hill AVS, Gilbert SC (2017) ChAdOx1 and MVA based vaccine candidates against MERS-CoV elicit neutralising antibodies and cellular immune responses in mice. Vaccine 35(30):3780–3788

Ashour MH, Elkhatib FW, Rahman MM, Elshabrawy AH (2020) Insights into the recent 2019 novel coronavirus (SARS-CoV-2) in light of past human coronavirus outbreaks. Pathogens 9(3). https://doi.org/10.3390/pathogens9030186

Bai Y, Yao L, Wei T, Tian F, Jin D-Y, Chen L, Wang M (2020) Presumed asymptomatic carrier transmission of COVID-19. JAMA. https://doi.org/10.1001/jama.2020.2565

Buchholz UJ, Bukreyev A, Yang L, Lamirande EW, Murphy BR, Subbarao K, Collins PL (2004) Contributions of the structural proteins of severe acute respiratory syndrome coronavirus to protective immunity. Proc Natl Acad Sci U S A 101(26):9804–9809

Channappanavar R, Fett C, Zhao J, Meyerholz DK, Perlman S (2014) Virus-specific memory CD8 T cells provide substantial protection from lethal severe acute respiratory syndrome coronavirus infection. J Virol 88(19):11034

van Doremalen N, Bushmaker T, Morris DH, Holbrook MG, Gamble A, Williamson BN, Tamin A, Harcourt JL, Thornburg NJ, Gerber SI, Lloyd-Smith JO, de Wit E, Munster VJ (2020) Aerosol and surface stability of SARS-CoV-2 as compared with SARS-CoV-1. N Engl J Med. https://doi.org/10.1056/NEJMc2004973

Dresden D (2020) Coronavirus vaccine: everything you need to know. 12 March 2020. Available from https://www.medicalnewstoday.com/articles/coronavirus-vaccine

Fehr AR, Perlman S (2015) Coronaviruses: an overview of their replication and pathogenesis. In: Maier HJ, Bickerton E, Britton P (eds) Coronaviruses: methods and protocols. Springer New York, New York, pp 1–23

Forni D, Cagliani R, Clerici M, Sironi M (2017) Molecular evolution of human coronavirus genomes. Trends Microbiol 25(1):35–48

Fulker R, Wasmoen T, Atchison R, Chu HJ, Acree W (1995) Efficacy of an inactivated vaccine against clinical disease caused by canine coronavirus. In: Talbot PJ, Levy GA (eds) Corona- and related viruses: current concepts in molecular biology and pathogenesis. Springer US, Boston, MA, pp 229–234

Gao W, Tamin A, Soloff A, D'Aiuto L, Nwanegbo E, Robbins PD, Bellini WJ, Barratt-Boyes S, Gambotto A (2003) Effects of a SARS-associated coronavirus vaccine in monkeys. Lancet 362 (9399):1895–1896

Ji W, Wang W, Zhao X, Zai J, Li X (2020) Cross-species transmission of the newly identified coronavirus 2019-nCoV. J Med Virol 92(4):433–440

Kumar S, Maurya VK, Prasad AK, Bhatt MLB, Saxena SK (2020) Structural, glycosylation and antigenic variation between 2019 novel coronavirus (2019-nCoV) and SARS coronavirus (SARS-CoV). Virus Dis 2020:1–9. https://doi.org/10.1007/s13337-020-00571-5

Li C K-f, Wu H, Yean H, Ma S, Wang L, Zhang M, Tang X, Temperton NJ, Weiss RA, Brenchley JM, Douek DC, Mongkolsapaya J, Tran B-H, Lin C-l S, Screaton GR, Hou J-l, McMichael AJ, Xu X-N (2008) T cell responses to whole SARS coronavirus in humans. J Immunol 181(8):5490

Liu C, Zhou Q, Li Y, Garner LV, Watkins SP, Carter LJ, Smoot J, Gregg AC, Daniels AD, Jervey S, Albaiu D (2020) Research and development on therapeutic agents and vaccines for COVID-19 and related human coronavirus diseases. ACS Cent Sci 6(3):315–331

Menachery VD, Yount BL, Sims AC, Debbink K, Agnihothram SS, Gralinski LE, Graham RL, Scobey T, Plante JA, Royal SR, Swanstrom J, Sheahan TP, Pickles RJ, Corti D, Randell SH, Lanzavecchia A, Marasco WA, Baric RS (2016) SARS-like WIV1-CoV poised for human emergence. Proc Natl Acad Sci U S A 113(11):3048

Menachery VD, Gralinski LE, Mitchell HD, Dinnon KH, Leist SR, Yount BL, McAnarney ET, Graham RL, Waters KM, Baric RS (2018) Combination attenuation offers strategy for live attenuated coronavirus vaccines. J Virol 92(17):e00710–e00718

Narayanan K, Ramirez SI, Lokugamage KG, Makino S (2015) Coronavirus nonstructural protein 1: common and distinct functions in the regulation of host and viral gene expression. Virus Res 202:89–100

Shen X, Sabir JSM, Irwin DM, Shen Y (2019) Vaccine against Middle East respiratory syndrome coronavirus. Lancet Infect Dis 19(10):1053–1054

Stockman LJ, Bellamy R, Garner P (2006) SARS: systematic review of treatment effects. PLoS Med 3(9):e343

Takasuka N, Fujii H, Takahashi Y, Kasai M, Morikawa S, Itamura S, Ishii K, Sakaguchi M, Ohnishi K, Ohshima M, Hashimoto S-i, Odagiri T, Tashiro M, Yoshikura H, Takemori T, Tsunetsugu-Yokota Y (2004) A subcutaneously injected UV-inactivated SARS coronavirus vaccine elicits systemic humoral immunity in mice. Int Immunol 16(10):1423–1430

Tekes G, Thiel HJ (2016) Chapter 6—Feline coronaviruses: pathogenesis of feline infectious peritonitis. In: Ziebuhr J (ed) Advances in virus research, vol 96. Academic Press, Cambridge, MA, pp 193–218

Tripet B, Howard MW, Jobling M, Holmes RK, Holmes KV, Hodges RS (2004) Structural characterization of the SARS-coronavirus spike S fusion protein core. J Biol Chem 279 (20):20836–20849

Tseng C-T, Sbrana E, Iwata-Yoshikawa N, Newman PC, Garron T, Atmar RL, Peters CJ, Couch RB (2012) Immunization with SARS coronavirus vaccines leads to pulmonary immunopathology on challenge with the SARS virus. PLoS One 7(4):e35421

Wan XF, Ataman D, Xu D (2005) Application of computational biology in understanding emerging infectious diseases: Inferring the biological function for S-M complex of SARS-CoV. Prog Bioinformatics. Nova Science Publishers, New York, pp 55–80

Wu A, Peng Y, Huang B, Ding X, Wang X, Niu P, Meng J, Zhu Z, Zhang Z, Wang J, Sheng J, Quan L, Xia Z, Tan W, Cheng G, Jiang T (2020) Genome composition and divergence of the novel coronavirus (2019-nCoV) originating in China. Cell Host Microbe 27(3):325–328

Yuen K-Y, Wong SSY, Peiris JSM (2007) The severe acute respiratory syndrome. In: Fong IW, Alibeck K (eds) New and evolving infections of the 21st century. Springer New York, New York, pp 163–193

Zhou J, Wang W, Zhong Q, Hou W, Yang Z, Xiao S-Y, Zhu R, Tang Z, Wang Y, Xian Q, Tang H, Wen L (2005) Immunogenicity, safety, and protective efficacy of an inactivated SARS-associated coronavirus vaccine in rhesus monkeys. Vaccine 23(24):3202–3209

Zhou F, Yu T, Du R, Fan G, Liu Y, Liu Z, Xiang J, Wang Y, Song B, Gu X, Guan L, Wei Y, Li H, Wu X, Xu J, Tu S, Zhang Y, Chen H, Cao B (2020) Clinical course and risk factors for mortality of adult inpatients with COVID-19 in Wuhan, China: a retrospective cohort study. Lancet 395 (10229):1054–1062

Zou L, Ruan F, Huang M, Liang L, Huang H, Hong Z, Yu J, Kang M, Song Y, Xia J, Guo Q, Song T, He J, Yen H-L, Peiris M, Wu J (2020) SARS-CoV-2 viral load in upper respiratory specimens of infected patients. N Engl J Med 382(12):1177–1179

Chapter 12
Classical Coronaviruses

Nitesh Kumar Jaiswal and Shailendra K. Saxena

Abstract In the last week of December 2019, few patients with the history of pyrexia of unknown origin and symptoms of lower respiratory tract infections were detected in Wuhan, a well-known area as the largest metropolitan city located in the province of Hubei, China. On further investigation, a novel coronavirus was identified as the causative pathogen, which later on provisionally named as 2019 novel coronavirus (2019-nCoV). Coronaviruses are predominantly found in warm-blooded animals and birds and cause various respiratory complications and multiorgan failure in the immunocompromised individuals. Human coronaviruses were first identified in 1965 and are responsible for the respiratory tract infections in major proportion of population worldwide; at least five new human coronaviruses have been identified, including severe acute respiratory syndrome coronavirus (SARS-CoV) in 2002–2003 and Middle East respiratory syndrome coronavirus (MERS-CoV) in 2012. The background related to the origin and classification of coronaviruses is reviewed here.

Keywords 2019-nCoV · MERS-CoV · SARS-CoV · Strain 229E · Strain OC43 · Coronavirus

---

Nitesh Kumar Jaiswal and Shailendra K. Saxena contributed equally as first author.

N. K. Jaiswal (✉)
Department of Microbiology, Zydus Medical College and Hospital, Dahod, Gujarat, India

S. K. Saxena
Centre for Advanced Research (CFAR), Faculty of Medicine, King George's Medical University (KGMU), Lucknow, India
e-mail: shailen@kgmcindia.edu

S. K. Saxena (ed.), *Coronavirus Disease 2019 (COVID-19)*, Medical Virology: from Pathogenesis to Disease Control, https://doi.org/10.1007/978-981-15-4814-7_12

12.1 Introduction

Coronaviruses isolated from several species consist of a group of large, enveloped, single plus stranded RNA viruses, and these viruses are earlier known to cause acute rhinitis and diarrhea in humans (Drosten et al. 2003a; Chen et al. 2020). A human coronavirus named SARS-CoV (severe acute respiratory syndrome coronavirus) was associated with the SARS outbreak in the year 2002–2003 (Zhong et al. 2003; Drosten et al. 2003b; Fouchier et al. 2003; Ksiazek et al. 2003). Similarly 10 year after SARS outbreak, another deadly human coronavirus [Middle East respiratory syndrome coronavirus (MERS-CoV)] emerged in the Middle East nations (Zaki et al. 2012). Recently, a new coronavirus named 2019-nCoV (belonging to the family *Coronaviridae* and subfamily Orthocoronavirinae) has erupted in the region of Wuhan (China) with severe respiratory tract infections in human. This virus is distinct from MERS-CoV and SARS-CoV; animal-to-human transmission has been considered as the origin of this outbreak, as most of the patients had a history to visit a local fish and wild animal market in Wuhan during the epidemic (Chan et al. 2020; Huang et al. 2020; Zhu et al. 2020). Animal-to-human and inter-human transmissions of this viral infection through respiratory route were established by certain group of scientists (Lu et al. 2020; Ji et al. 2020). Patient isolation and accurate timely diagnosis are the hallmarks to control this new epidemic. It is also important to acquire knowledge from the history and evolution of coronaviruses. Phylogenetic analysis and its relation with different natural hosts of these viruses can help one to estimate its genetic variability which in turn has important applications for the viral etiopathogenesis, clinical manifestations, and vaccine development.

12.2 Origin and Evolution of Coronaviruses

In the mid-1930s, a severe respiratory infection of chicken was considered to be the earliest known disease by coronaviruses, infection presently known as avian infectious bronchitis, caused by avian infectious bronchitis virus (IBV). The era of human coronaviruses began in the year 1965 when Tyrrell and Bynoe observed that they could serially sub-passage a virus in tissue culture while doing research on human participants at the Common Cold Unit close to Salisbury, UK; they named this virus as B814 (Tyrrell and Bynoe 1966). Their experiment demonstrated that common colds could be transmitted by nasal secretion that did not contain rhinoviruses, and further in vitro experiments of nasal swabs from these participants had been inoculated onto organ cultures obtained from respiratory tract cell lines. They discovered the presence of enveloped RNA viruses with the feature morphology of coronaviruses similar to the previously defined infectious bronchitis virus. They were unable to grow the virus in tissue culture at that time. Subsequently in the year 1966, Hamre and Procknow succeeded in growing a new virus (229E) of unexpected

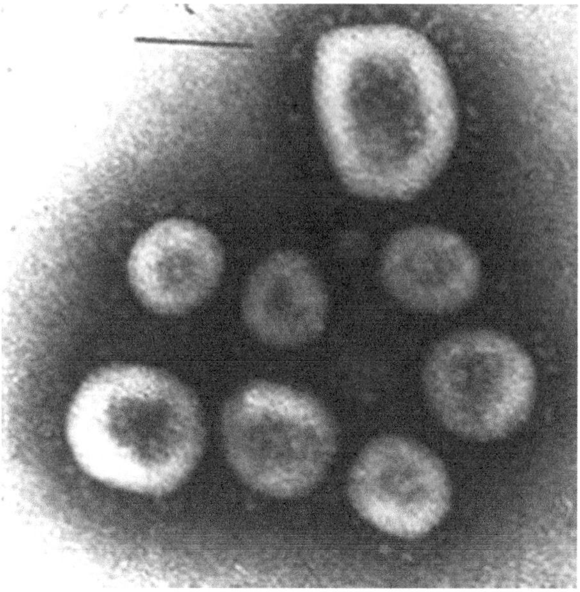

Fig. 12.1 Coronavirus OC16 (McIntosh et al. 1967a)

tissue culture properties from a medical graduate with symptoms of common cold (Hamre and Procknow 1966).

Both B814 and 229E were ether sensitive and required a lipid-containing coat for infectivity. These two viruses were unassociated with any known orthomyxoviruses. McIntosh et al. (1967a) had used the similar technique of Tyrrell and Bynoe to extract the multiple strains of ether-susceptible agents from human volunteers. These viruses were designated OC as they were grown in the organ culture of respiratory tract. In the year 1967, Almeida and Tyrrell were able to demonstrate the similar morphology of B814 and IBV under the electron microscopy of the fluid obtained from the inoculated organ culture. The virus particles were of size around 80–150 nm, pleomorphic, enveloped with membrane coating, and multiple club-shaped surface projections (Almeida and Tyrrell 1967). All of these (229E, OC, B814, and IBV) along with the virus causing mouse hepatitis and transmissible gastroenteritis infection of swine had a similar morphology under electron microscopy (McIntosh et al. 1967b; Witte et al. 1968) (Fig. 12.1). The new group of these viruses was named coronavirus in 1968 (corona means the crown-like structure with surface projection), reflecting their morphology in the electron microscope, and further Coronaviridae was accepted as their family name in 1975 (Tyrrell et al. 1975).

229E and OC 43 were the only two human coronaviruses discovered before the outbreak of SARS-CoV. Both of them were recognized as less pathogenic causing mild flu-like infection, thus not being explored further in the research field. SARS emerged in the late 2002 from South East Asia with an epicenter of China and spread expeditiously all over the world. Taking after the acknowledgement that SARS was caused by a new human coronavirus (SARS-CoV), it was found to be more pathogenic and causes serious respiratory complications (Li et al. 2005; Ren et al. 2008; Peiris et al. 2003). The genomic sequencing of SARS-CoV was comfortably achieved as it had the capacity to grow readily in tissue culture. Two other modern human coronaviruses, NL63 and HKU-1, were also found in association with respiratory disease.

Ten years after SARS, yet another profoundly pathogenic condition, Middle East respiratory syndrome or MERS appeared that took its heaviest toll on wealthy urban area of Middle East nations (Zaki et al. 2012). SARS-CoV and MERS-CoV were transmitted directly to people from feline animals (civet) and dromedary camels, respectively (Guan et al. 2003; Alagaili et al. 2014; Hemida et al. 2013). Extensive research studies of these two viruses have helped to know about the biological properties and exceptional perception of human coronaviruses.

In late December 2019, few patients with the history of unexplained fever and symptoms of lower respiratory tract infections were detected in Wuhan, an area avowed as the largest metropolitan city located in the Hubei province of China. The etiology of this unknown respiratory infection was not being able to be established; it was classified as pneumonia of unknown etiology initially. Further investigation by the local authority of Chinese Center for Disease Control and Prevention identified the causative pathogen and provisionally named 2019 novel coronavirus (2019-nCoV). On February 11, 2020, the WHO Director-General, Dr. Tedros Adhanom Ghebreyesus, announced that the disease caused by this new coronavirus is COVID-19 which is the acronym of "coronavirus disease 2019." The WHO raised the threat to this CoV epidemic in the category of "very high" level in February 2020. Various organizations throughout the globe are working to establish countermeasures to prevent possible devastating effects from this new human coronavirus.

The reestablished intrigued in this bunch of infections has driven to the disclosure of a plenty of coronaviruses and their capacity to jump over different animal species over a period of time.

12.3 Classification

The International Committee on Taxonomy (ICT) of viruses has classified coronaviruses under family *Coronaviridae* and order *Nidovirales*. Toroviruses and coronaviruses are the two representative genera of the family *Coronaviridae*; they have been further classified in subfamily *Coronavirinae*. On the basis of rooted and unrooted genetic trees and partial nucleotide sequence of RNA-dependent RNA polymerase, subfamily *Coronavirinae* has been recognized and classified into four genera—alpha (α) coronavirus, beta (β) coronavirus, gamma (γ) coronavirus, and delta (δ) coronavirus (Woo et al. 2009, 2012a). The α coronaviruses and β coronaviruses infect only warm-blooded animals. The γ coronaviruses and δ coronaviruses infect birds, but some of them can also infect mammals. Phylogenetic relationship of various human and animal coronaviruses and the list of known

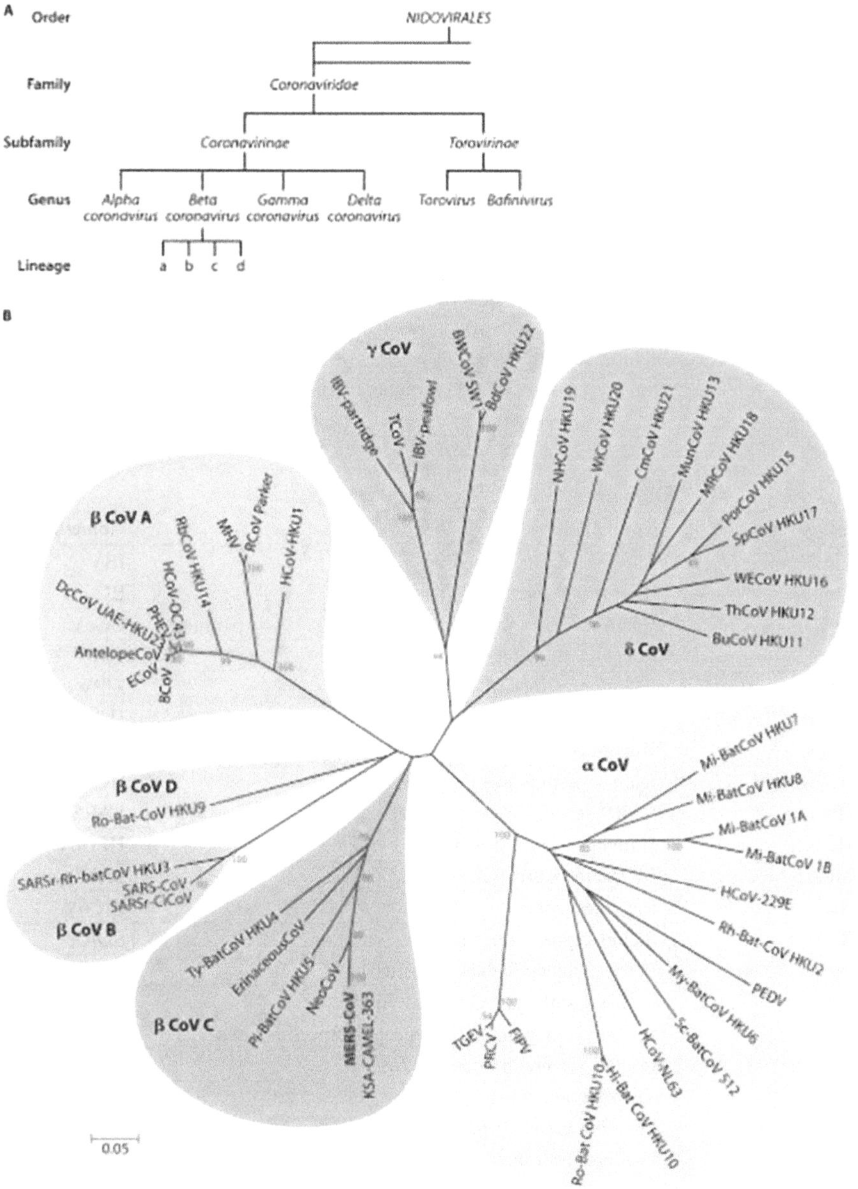

Fig. 12.2 Phylogenetic classification of coronaviruses. (Source: Cui J., Lei F and Shi ZL. Review: Origin and evolution of pathogenic coronaviruses. Nature Reviews Microbiology, December 2018)

coronaviruses are shown in Fig. 12.2 and Tables 12.1, 12.2, and 12.3, respectively. Infection with α coronaviruses and β coronaviruses leads to respiratory disorders in human and gastrointestinal disorders in animals (Su et al. 2016; Forni et al. 2017; Gorbalenya et al. 2006). Currently available genetic sequence databases of all human

Table 12.1 Coronaviruses under different genera

Genus	Species
Alpha coronavirus	Human coronavirus 229E, human coronavirus NL63, Miniopterus bat coronavirus 1, Miniopterus bat coronavirus HKU8, porcine epidemic diarrhea virus, Rhinolophus bat coronavirus HKU2, Scotophilus bat coronavirus 512
Beta coronavirus	Beta coronavirus 1, human coronavirus HKU1, murine coronavirus, Pipistrellus bat coronavirus HKU5, Rousettus bat coronavirus HKU9, Severe acute respiratory syndrome-related coronavirus, Severe acute respiratory syndrome coronavirus 2, Tylonycteris bat coronavirus HKU4, Middle East respiratory syndrome-related coronavirus, human coronavirus OC43, hedgehog coronavirus 1 (EriCoV)
Gamma coronavirus	Infectious bronchitis virus (IBV), beluga whale coronavirus SW1, infectious bronchitis virus
Delta coronavirus	Bulbul coronavirus HKU11, porcine coronavirus HKU15

Table 12.2 Coronaviruses and their different natural hosts

Natural host	Coronaviruses	Abbreviation
Chicken	Infectious bronchitis virus	IBV
Cattle	Bovine coronavirus	BCoV
Dog	Canine enteric coronavirus	CCoV
Cat	Feline coronavirus	FCoV
Cat	Feline infectious peritonitis virus	FIPV
Humans	Human coronavirus 229E	HCoV-229E
Humans	Human coronavirus NL63	HCoV-NL63
Humans	Human coronavirus OC43	HCoV-OC43
Humans	SARS-coronavirus	SARS-CoV
Humans	Human enteric coronavirus	HECoV
Mouse	Murine hepatitis virus	MHV
Rat	Rat coronavirus	RtCoV
Rat	Sialodacryoadenitis virus	SDAV
Pig	Porcine epidemic diarrhea virus	PEDV
Pig	Transmissible gastroenteritis virus	TGEV
Pig	Porcine hemagglutinating encephalomyelitis virus	HEV
Pig	Porcine respiratory coronavirus	PRCoV
Turkey	Turkey coronavirus	TCoV
Pheasant	Pheasant coronavirus	PhCoV
Shearwater	Puffinosis coronavirus	PCoV

coronaviruses reveal with its animal origins: Viruses of the family *Coronaviridae* are listed, with their abbreviations and natural animal reservoir (Cavanagh 1997) (Table 12.2). Domestic animals can act as an intermediate host to transmit these viruses from their reservoir host to the humans. Sometimes domestic animal may acquire infection with closely related zoonotic coronaviruses (Woo et al. 2009, 2012a). Extensive research study on the source of infection of SARS-CoV and

Table 12.3 List of human and animal coronaviruses

Human coronaviruses	Animal coronaviruses
HCoV-229E (human coronavirus 229E)	Antelope CoV (Sable antelope coronaviruses)
HCoV-HKU1 (human coronavirus HKU1)	BCoV (bovine coronaviruses)
HCoV-NL63 (human coronavirus NL63)	BdCoV HKU22 (bottlenose dolphin coronavirus HKU22)
HCoV-OC43 (human coronavirus OC43)	BuCoV HKU11 (bulbul coronavirus HKU11)
SARS-CoV (SARS coronavirus)	BWCoV-SW1 (beluga whale coronavirus SW1)
MERS-CoV (Middle East respiratory syndrome coronavirus)	CMCoV HKU21 (common-moorhen coronavirus HKU21)
	DcCoV UAE-HKU23 (dromedary camel coronavirus UAE-HKU23)
2019-nCoV (COVID 19)	ECoV (equine coronavirus)
	ErinaceousCoV (beta coronavirus *Erinaceus*)
	FIPV (feline infectious peritonitis virus)
	Hi-BatCoV HKU10 (*Hipposideros* bat coronavirus HKU10)
	IBV-partridge (avian infectious bronchitis virus partridge isolate)
	IBV-peafowl (avian infectious bronchitis virus peafowl isolate)
	KSA-CAMEL-363 (KSA-CAMEL-363 isolate of Middle East respiratory syndrome coronavirus)
	MHV (murine hepatitis virus)
	Mi-BatCoV 1A (*Miniopterus* bat coronavirus)
	Mi-BatCoV 1B (*Miniopterus* bat coronavirus 1B)
	Mi-BatCoV HKU7 (*Miniopterus* bat coronavirus HKU7)
	Mi-BatCoV HKU8 (*Miniopterus* bat coronavirus HKU8)
	MRCoV HKU18 (magpie robin coronavirus HKU18)
	MunCoV HKU13 (munia coronavirus HKU13)
	My-BatCoV HKU6 (*Myotis* bat coronavirus HKU6)
	NeoCoV (coronavirus *Neoromicia*)
	NHCoV HKU19 (night heron coronavirus)
	PEDV (porcine epidemic diarrhea virus)
	PHEV (porcine hemagglutinating encephalomyelitis virus)
	Pi-BatCoV-HKU5 (*Pipistrellus* bat coronavirus HKU5)
	PorCoV HKU15 (porcine coronavirus HKU15)
	PRCV (porcine respiratory coronavirus)
	RbCoV HKU14 (rabbit coronavirus HKU14)
	RCoV parker (rat coronavirus Parker)
	Rh-BatCoV HKU2 (*Rhinolophus* bat coronavirus HKU2)
	Ro-BatCoV-HKU9 (*Rousettus* bat coronavirusHKU9)

(continued)

Table 12.3 (continued)

Human coronaviruses	Animal coronaviruses
	Ro-BatCoV HKU10 (*Rousettus* bat coronavirus HKU10)
	SARSr-CiCoV (SARS-related palm civet coronavirus)
	SARSr-Rh-BatCoV HKU3 (SARS-related *Rhinolophus* bat coronavirus HKU3)
	Sc-BatCoV 512 (*Scotophilus* bat coronavirus 512)
	SpCoV HKU17 (sparrow coronavirus HKU17)
	TCoV (turkey coronavirus)
	TGEV (transmissible gastroenteritis virus)
	ThCoV HKU12 (thrush coronavirus HKU12)
	Ty-BatCoV-HKU4 (*Tylonycteris* bat coronavirus HKU4)
	WECoV HKU16 (white-eye coronavirus HKU16)
	WiCoV HKU20 (wigeon coronavirus HKU20)

MERS-CoV has found bat as a reservoir, and it has laid to the better understanding of coronavirus microbiology. At present, seven species of α coronavirus and four species of β coronavirus have been identified only in bats (Su et al. 2016; Lau et al. 2013; Perlman and Netland 2009; Woo et al. 2012b; Graham et al. 2013; Hu et al. 2015; de Wit et al. 2016; Wang et al. 2018; Lin et al. 2017).

12.4 Future Perspectives

As we deal with the thriving pandemic of novel highly contagious human coronavirus disease (Covid-19), caused by the severe acute respiratory syndrome coronavirus 2 (SARS-CoV-2), the human race is locked in a battle against it. Genomic sequences of closely related viruses silently circulate in bats. The molecular epidemiological information reveals that a virus of bat origin infecting unidentified animal species sold in Wuhan sea food markets could be the possible source of this infection. Several research studies have focused on the evaluation of the zoonotic potentials of animal coronaviruses since the outbreak of SARS-CoV and MERS-CoV. The high diversity of coronaviruses detected in bats and their genomic variability increases the risk of interspecies transmission. 2019-nCoV highlights the importance of bats as a reservoir for new viruses causing infection in humans, and it can also be used as a good model to design studies and strategies to prevent future emergence of new zoonotic diseases. It is important to increase the efforts to characterize viral genome of different animal coronaviruses and also to look for the viral evolution and adaptation to their natural hosts. The possibility to predict the interspecies transmission can be helpful in planning of specific surveillance programs and act quickly in an outbreak. Therefore, extensive research studies in finding experimental models for the emerging zoonotic viral diseases are absolutely mandatory.

Executive Summary

- Classical coronaviruses consist of a group of large, enveloped, RNA viruses.
- Recently, a new coronavirus named 2019-nCoV (belonging to the family *Coronaviridae* and subfamily *Orthocoronavirinae*) has erupted in the South East Asian country.
- Subfamily *Coronavirinae* consists of four genera—alpha coronavirus, beta coronavirus, gamma coronavirus, and delta coronavirus.
- Alpha coronaviruses and beta coronaviruses cause respiratory disorders in human and gastrointestinal disorders in animals.
- Animal-to-human and inter-human transmissions are frequently seen in theses group of viruses.

References

Alagaili AN et al (2014) Middle East respiratory syndrome coronavirus infection in dromedary camels in Saudi Arabia. MBio 5:e00884–e00814

Almeida JD, Tyrrell DA (1967) The morphology of three previously uncharacterized human respiratory viruses that grow in organ culture. J Gen Virol 1:175–178

Cavanagh D (1997) Nidovirales: a new order comprising Coronaviridae and Arteriviridae. Arch Virol 14:629–633

Chan JF, Yuan S, Kok KH et al (2020) A familial cluster of pneumonia associated with the 2019 novel coronavirus indicating person to person transmission: a study of a family cluster. Lancet. https://doi.org/10.1016/S0140/6736(20)30154-9

Chen Y, Liu Q, Guo D (2020) Emerging coronaviruses: genome structure, replication, and pathogenesis. J Med Virol. https://doi.org/10.1002/jmv.25681

Drosten C, Günther S, Preiser W et al (2003a) Identification of a novel coronavirus associated with severe acute respiratory syndrome. N Engl J Med 348:1967–1976

Drosten C et al (2003b) Identification of a novel coronavirus in patients with severe acute respiratory syndrome. N Engl J Med 348:1967–1976

Forni D, Cagliani R, Clerici M, Sironi M (2017) Molecular evolution of human coronavirus genomes. Trends Microbiol 25:35–48

Fouchier RA et al (2003) Aetiology: Koch's postulates fulfilled for SARS virus. Nature 423:240

Gorbalenya AE et al (2006) Nidovirales: evolving the largest RNA virus genome. Virus Res 117:17–37

Graham RL, Donaldson EF, Baric RS (2013) A decade after SARS: strategies for controlling emerging coronaviruses. Nat Rev Microbiol 11:836–848

Guan Y et al (2003) Isolation and characterization of viruses related to the SARS coronavirus from animals in southern China. Science 302:276–278

Hamre D, Procknow JJ (1966) A new virus isolated from the human respiratory tract. Proc Soc Exp Biol Med 121:190–193

Hemida MG et al (2013) Middle East Respiratory Syndrome (MERS) coronavirus seroprevalence in domestic livestock in Saudi Arabia, 2010 to 2013. Euro Surveill 18:21–27

Hu B, Ge X, Wang LF, Shi Z (2015) Bat origin of human coronaviruses. Virol J 12:221

Huang C, Wang Y, Li X et al (2020) Clinical features of patients infected with 2019 novel coronavirus in Wuhan, China. Lancet. https://doi.org/10.1016/S0140-6736(20)30183-5

Ji W, Wang W, Zhao X, Zai J, Li X (2020) Homologous recombination within the spike glycoprotein of the newly identified coronavirus may boost cross-species transmission from snake to human. J Med Virol. https://doi.org/10.1002/jmv.25682

Ksiazek TG et al (2003) A novel coronavirus associated with severe acute respiratory syndrome. N Engl J Med 348:1953–1966

Lau SK et al (2013) Genetic characterization of Betacoronavirus lineage C viruses in bats reveals marked sequence divergence in the spike protein of pipistrellus bat coronavirus HKU5 in Japanese pipistrelle: implications for the origin of the novel Middle East respiratory syndrome coronavirus. J Virol 87:8638–8650

Li F et al (2005) Structure of SARS coronavirus spike receptor-binding domain complexed with receptor. Science 309:1864–1868

Lin XD et al (2017) Extensive diversity of coronaviruses in bats from China. Virology 507:1–10

Lu H, Stratton CW, Tang YW (2020) Outbreak of pneumonia of unknown etiology in Wuhan China: the mystery and the miracle. J Med Virol. https://doi.org/10.1002/jmv.25678

McIntosh K, Dees JH, Becker WB, Kapikian AZ, Chanock RM (1967a) Recovery in tracheal organ cultures of novel viruses from patients with respiratory disease. Proc Natl Acad Sci U S A 57:933–940

McIntosh K, Becker WB, Chanock RM (1967b) Growth in suckling-mouse brain of "IBV-like" viruses from patients with upper respiratory tract disease. Proc Natl Acad Sci U S A 58:2268–2273

Peiris JS, Lai ST, Poon LL et al (2003) Coronavirus as a possible cause of severe acute respiratory syndrome. Lancet 361:1319–1325

Perlman S, Netland J (2009) Coronaviruses post-SARS: update on replication and pathogenesis. Nat Rev Microbiol 7:439–450

Ren W et al (2008) Difference in receptor usage between severe acute respiratory syndrome (SARS) coronavirus and SARS-like coronavirus of bat origin. J Virol 82:1899–1907

Su S et al (2016) Epidemiology, genetic recombination, and pathogenesis of coronaviruses. Trends Microbiol 24:490–502

Tyrrell DA, Bynoe ML (1966) Cultivation of viruses from a high proportion of patients with colds. Lancet 1:76–77

Tyrrell DA, Almeida JD, Cunningham CH et al (1975) Coronaviridae. Intervirology 5:76–82

Wang L, Su S, Bi Y, Wong G, Gao GF (2018) Bat—origin coronaviruses expand their host range to pigs. Trends Microbiol 26:466–470

de Wit E, van Doremalen N, Falzarano D, Munster VJ (2016) SARS and MERS: recent insights into emerging coronaviruses. Nat Rev Microbiol 14:523–534

Witte KH, Tajima M, Easterday BC (1968) Morphologic characteristics and nucleic acid type of transmissible gastroenteritis virus of pigs. Arch Gesamte Virusforsch 23:53–70

Woo PC et al (2009) Coronavirus diversity, phylogeny and interspecies jumping. Exp Biol Med 234:1117–1127

Woo PC et al (2012a) Discovery of seven novel mammalian and avian coronaviruses in the genus deltacoronavirus supports bat coronaviruses as the gene source of alphacoronavirus and betacoronavirus and avian coronaviruses as the gene source of gammacoronavirus and deltacoronavirus. J Virol 86:3995–4008

Woo PC et al (2012b) Discovery of seven novel mammalian and avian coronaviruses in the genus deltacoronavirus supports bat coronaviruses as the gene source of alphacoronavirus and betacoronavirus and aviancoronaviruses as the gene source of gammacoronavirus and deltacoronavirus. J Virol 86:3995–4008

Zaki AM, Boheemen S, Bestebroer TM, Osterhaus AD, Fouchier RA (2012) Isolation of anovel coronavirus from a man with pneumonia in Saudi Arabia. N Engl J Med 367:1814–1820

Zhong NS et al (2003) Epidemiology and cause of severe acute respiratory syndrome (SARS) in Guangdong, People's Republic of China, in February. Lancet 362:1353–1358

Zhu N, Zhang D, Wang W et al (2020) A novel coronavirus from patients with pneumonia in China, 2019. N Engl J Med. https://doi.org/10.1056/NEJMoa2001017

Chapter 13
Emergence and Reemergence of Severe Acute Respiratory Syndrome (SARS) Coronaviruses

Preeti Baxi and Shailendra K. Saxena

Abstract The positive-strand RNA viruses, severe acute respiratory syndrome coronavirus (SARS-CoV) and recently emerged COVID-19 epidemics, demonstrated the transmission capability of the coronaviruses by crossing the species barrier and emergence in humans. The source of coronavirus disease 2019 (COVID-19) is severe acute respiratory syndrome coronavirus 2 (SARS-CoV-2), firstly reported in December 2019 at Wuhan, China. COVID-19 is a kind of viral pneumonia. The outbreak of SARS-CoV-2 (COVID-19) has been reported as the introduction of the third highly pathogenic coronavirus which crossed the species barrier and spread into the human population. Severe acute respiratory syndrome coronavirus (SARS-CoV) and the Middle East respiratory syndrome coronavirus (MERS-CoV) were the first two epidemic viruses, respectively, in the twenty-first century. Introduction of the 2019 novel coronaviruses (2019-nCoV) in human population is a worldwide concern, and this might have generated via RNA recombination among the previous reported coronaviruses. The COVID-19 is spreading in an alarming rate, and till date no vaccine or specific medicines are available in the market. The newly emerged coronavirus COVID-19 is strongly related to SARS-CoV except little dissimilarity. In this chapter, we will discuss about the alterations and variations in antigenicity, structural changes, and RNA recombination which might be responsible for the COVID-19 emergence.

Keywords Coronavirus · COVID-19 · SARS-CoV · SARS-CoV-2 · Antigenicity · Glycosylation · Spike glycoprotein · RNA recombination

---

Preeti Baxi and Shailendra K. Saxena contributed equally as first author.

P. Baxi (✉)
Phytosanitory Laboratory, Department of Plant Molecular Biology and Biotechnology, Indira Gandhi Agriculture University, Raipur, India

S. K. Saxena
Centre for Advanced Research (CFAR), Faculty of Medicine, King George's Medical University (KGMU), Lucknow, India
e-mail: shailen@kgmcindia.edu

S. K. Saxena (ed.), *Coronavirus Disease 2019 (COVID-19)*, Medical Virology: from Pathogenesis to Disease Control, https://doi.org/10.1007/978-981-15-4814-7_13

Abbreviations

ACE2	Angiotensin-converting enzyme 2
DMV	Double membrane vesicle
HKU	Human coronavirus
IFN	Interferon
MERS	Middle East respiratory syndrome
NSP	Nonstructural protein
ORF	Open reading frame
RBD	Receptor-binding domain
RMSD	Root mean square deviation
SARS	Severe acute respiratory syndrome
ssRNA	Single-stranded ribonucleic acid

13.1 Introduction

Coronavirus belongs to the large family of viruses, i.e., *Coronaviridae* family and the order *Nidovirales* (Cui et al. 2019; Kumar et al. 2020; Phan et al. 2018). Genomic structures and phylogenetic relationship reveals that subfamily Coronavirinae contains the four genera alpha coronavirus and betacoronavirus which are restricted to mammals and responsible for respiratory illness in humans such as Middle East respiratory syndrome coronavirus (MERS-CoV) and SARS coronavirus (SARS-CoV) (Cui et al. 2019; Kumar et al. 2020; Payne 2017; Phan et al. 2018). The other two genera gamma coronavirus and delta coronavirus infect both birds and mammals (Cui et al. 2019; Kumar et al. 2020; Payne 2017; Phan et al. 2018). SARS-CoV and MERS-CoV show severe respiratory diseases in humans, and four others (HCoV-NL63, HCoV-229E, HCoV-OC43, and HKU1) cause normal upper respiratory illness in the hosts which are immunocompetent, even though some can induce severe infections in elders, young children, and infants (Cui et al. 2019; Cascella et al. 2020). Alpha and betacoronaviruses can give rise to intense burden on animals; these include recently emerged swine acute diarrhea syndrome coronavirus (SADS-CoV) and the porcine transmissible gastroenteritis virus and porcine enteric diarrhea virus (PEDV) (Cui et al. 2019; Banerjee et al. 2019).

Continuous development and urbanization increases the frequent integration of many different animals in crowded and populated places may have made easy the emergence and reemergence of a number of these viruses (Lau and Chan 2015; https://www.ncbi.nlm.nih.gov/books/NBK45714/). On the other hand, high mutation and recombination rates are reported in coronaviruses, which may allow the coronaviruses to cross the barrier of species and adopt to new hosts (Lau and Chan 2015; https://clarivate.com/wp-content/uploads/dlm_uploads/2020/01/CORONAVIRUS-REPORT-1.30.2020.pdf; Liu et al. 2020).

The 2003 SARS epidemic has awakened the world's researchers and scientists on the transmission capability of the coronaviruses from animals to humans. The

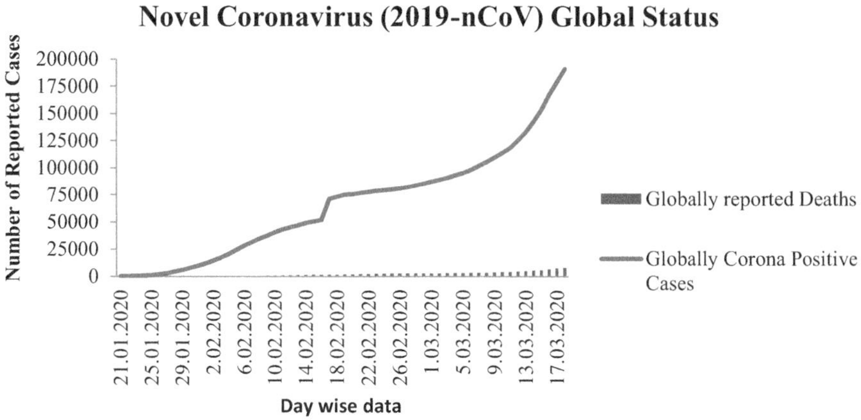

Fig. 13.1 Total reported confirmed cases and deaths due to novel coronavirus (2019-nCoV). (Data taken from https://www.who.int/emergencies/diseases/novel-coronavirus-2019/situation-reports)

primary reservoir of the SARS-CoV is horseshoe bat (Lau and Chan 2015; Wang et al. 2006). Current sequence databases revealed the animal origin of the human coronaviruses: MERS-CoV, SARS-CoV, HCoV-229E, and HCoV-NL63 are considered to have originated in bats (Cui et al. 2019; Wang et al. 2006). SARS-associated coronaviruses are found in bats from China and worldwide continuously (Lau and Chan 2015).

Recently in 2020, SARS-CoV-2, a novel coronavirus has emerged from China which affect globally with a total of 191,127 confirmed cases and 7807 deaths (as of March 18, 2020) (Kampf et al. 2020; https://www.who.int/docs/default-source/coronaviruse/situation-reports/20200314-sitrep-54-covid-19.pdf?sfvrsn=dcd46351_6). Figure 13.1 represents the total reported confirmed cases and deaths due to novel coronavirus (https://www.who.int/emergencies/diseases/novel-coronavirus-2019/situation-reports). The current pandemic of coronavirus-linked acute respiratory disease known as coronavirus disease 19 (COVID-19) is the third recognized spillover to humans through animal coronavirus in only the last two decades (https://doi.org/10.1038/s41564-020-0695-z; Kampf et al. 2020). Table 13.1 represents the discovery of human coronaviruses.

The viruses of the family Coronaviridae contain enveloped and positive-stranded RNA (Li et al. 2020; Coutard et al. 2020). Coronaviruses have the largest 26–32 kilobase (kb) positive-sense RNA genome (Li et al. 2020; Kumar et al. 2020; Shoeman and Fielding 2019). The four major structural proteins [membrane (M), envelope (E), spike (S), and nucleocapsid (N)] which create complete virus particle, encoded by the genome (Kumar et al. 2020; Wu et al. 2020; Guo et al. 2020).

The evolution and emergence of novel coronaviruses are mostly due to the lack of proofreading mechanism in RNA recombination among the present coronaviruses (Kumar et al. 2020). The S gene which encodes viral spike (S) glycoprotein has been proposed the higher frequency of recombination (Kumar et al. 2020). The novel coronavirus is closely related to bat SARS like betacoronavirus. Although, the

Table 13.1 Recent discoveries of human coronaviruses

Virus	*Coronaviriniae* genus	Cellular receptor	Host	Location	Year	References
HCoV-229E	Alphacoronavirus	Human aminopeptidase N (CD13)	Bats	–	1966	Lim et al. (2016), Kahn (2005), Cui et al. (2019), Human Coronavirus Types (2020); https://www.cdc.gov/coronavirus/types.html; Clarivate Analytics (2020)
HCoV-OC43	Betacoronavirus	9-*O*-Acetylated sialic acid	Cattle	–	1967	Lim et al. (2016), Kahn (2005), Cui et al. (2019), Hulswit et al. (2019), Human Coronavirus Types (2020); https://www.cdc.gov/coronavirus/types.html; Clarivate Analytics (2020)
SARS-CoV	Betacoronavirus	ACE2	Palm civets, bats	China	2003	Lim et al. (2016); https://www.who.int/ith/diseases/sars/en/; Cui et al. (2019), Adnan Shereen et al. (2020), Parrish et al. (2008), Human Coronavirus Types (2020); https://www.cdc.gov/coronavirus/types.html; Clarivate Analytics (2020)
HCoV-NL63	Alphacoronavirus	ACE2	Palm civets, bats	The Netherlands	2004	Lim et al. (2016), Cui et al. (2019); Rasool and Fielding (2010), Human Coronavirus Types (2020); https://www.cdc.gov/coronavirus/types.html; Clarivate Analytics (2020)
HKU1	Betacoronavirus	9-*O*-Acetylated sialic acid	Mice	Hong Kong	2005	Lim et al. (2016), Cui et al. (2019), Hulswit et al. (2019), Human Coronavirus Types (2020); https://www.cdc.gov/coronavirus/types.html; Clarivate Analytics (2020)
MERS-CoV (Middle East respiratory syndrome)	Betacoronavirus	DPP4	Bats, camels	Saudi Arabia	2012	Lim et al. (2016), Cui et al. (2019); Human Coronavirus Types (2020); https://www.cdc.gov/coronavirus/types.html; National Foundation for Infectious Diseases (2020) Coronavirus. https://www.nfid.org/infectious-diseases/coronaviruses/; Clarivate Analytics (2020)
COVID-19	Betacoronavirus	ACE2		China	2019	Lim et al. (2016), Kumar et al. (2020), Adnan Shereen et al. (2020), Tai et al. (2020), Human Coronavirus Types (2020); https://www.cdc.gov/coronavirus/types.html; Clarivate Analytics (2020)

longer spike protein of 2019-nCoV has remarkable difference when compared with the bat SARS-Cov and SARS-like coronaviruses (Kumar et al. 2020). In this chapter, we will discuss about the emergence and reemergence of the SARS coronaviruses and genetic variations from SARS-CoV to SARS-CoV-2.

13.1.1 SARS-CoV to SARS-CoV-2: RNA Recombination, Antigenic Shift, and Antigenic Drift

Reported phylogenetic analysis revealed the close relationship between COVID-19 and Bat SARS-like coronavirus (Lu et al. 2020; Kumar et al. 2020). Although both emerged from the SARS coronavirus, suggesting the newly spread 2019-nCov into human is very much related to SARS-CoV (Kumar et al. 2020). Figure 13.2 represents the evolutionary relationship between different lineage of betacoronavirus and their respective cellular receptors through which they interact with the host cell. As per sequence alignment data, the sequence of spike glycoprotein of COVID-19 and SARS-CoV shows 76.2% individuality, 87.2% resemblance, and 2% gaps. This data indicates that COVID-19 spike glycoprotein exhibits higher sequence similarity with 12.8% of variation with SARS-CoV. The study performed by Kumar et al. (2020), on variation of sequences of minimal receptor-binding domain (RBD), revealed 73.3% uniqueness, 83.9% matches, and 0.4% gaps, suggesting a difference of 16.1% and tertiary structure of minimal receptor-binding domain. COVID-19 may have changes in virus-binding capacity shift and infectivity into the receptors of the host cell, due to notable deviation in minimal RBD of S-glycoprotein.

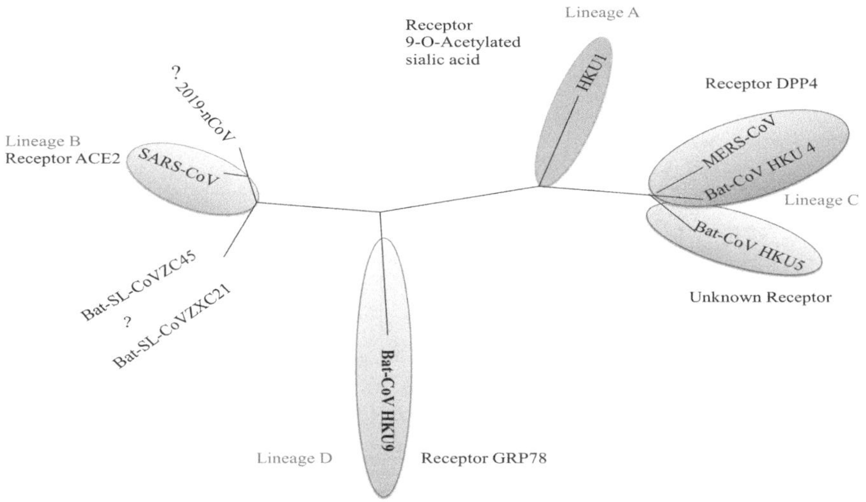

Fig. 13.2 Evolutionary relationship between different lineage of betacoronavirus and their respective receptor

13.2 Structure of Genome

The viruses belong to *Coronaviridae* family enclosed envelope and positive-stranded RNA (Li et al. 2020). Coronaviruses have the largest 26–32 kilobase (kb) positive-sense RNA genome. The genome of coronaviruses contains open reading frames (ORFs) of variable number (6–11). In the first ORF, two thirds of the total viral RNA are located. Upon infection, two large precursor polyproteins, namely pp1a and pp1ab, has been translated by viral genome in the host cells. These translated precursor polyproteins get processed into 16 mature nonstructural proteins (nsp1–nsp16) by viral proteinases. These nonstructural proteins have been numbered according to the position from N to C terminus (Kumar et al. 2020; Narayanan et al. 2014). Figure 13.3 depicts the structure of coronavirus representing essential proteins, accessory proteins, and nsp1–16. The remaining ORFs encode accessory proteins. The remaining part of viral genome encodes spike (S) glycoprotein, small envelope (E) protein, matrix (M) protein, and nucleocapsid protein (N). All these four are essential structural proteins. S protein is attached to the host receptor ACE2, including two subunits S1 and S2. The cellular tropism by RBD and virus host range is determined by S1. Virus cell membrane fusion is determined by S2 (Guo et al. 2020; Pradhan et al. 2020; Belouzard et al. 2009; Du et al. 2009).

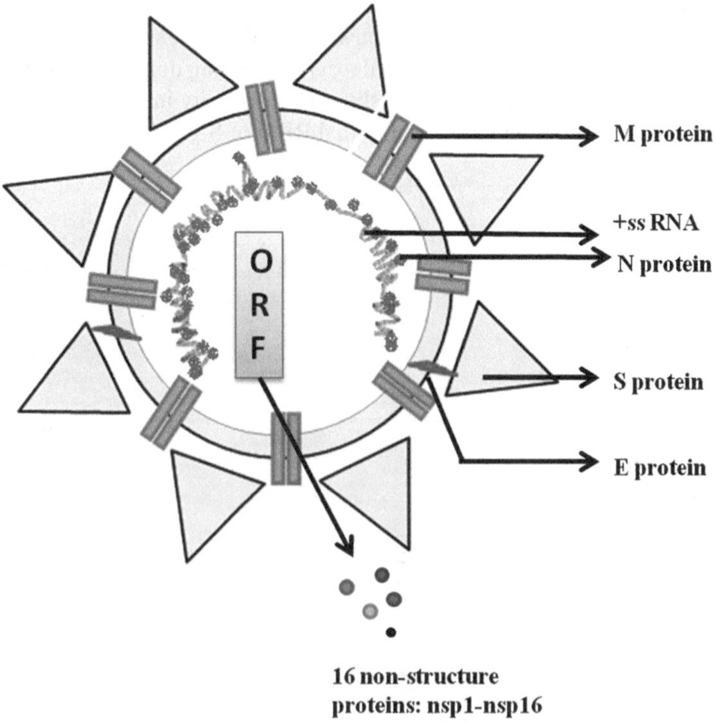

Fig. 13.3 Structure of coronavirus causing severe acute respiratory syndrome

Transmembrane nutrient transport and formation of envelop were performed by M protein. N protein and E protein along with several accessory proteins obstruct with host immune response (Guo et al. 2020; Pradhan et al. 2020). The nonstructural proteins play a vital function at the time of viral RNA replication and transcription (Kumar et al. 2020; Narayanan et al. 2014). The assigned functions of these 16 mature nonstructural proteins are mentioned in Table 13.2. Several nsps are

Table 13.2 Functions of 16 mature nonstructural proteins which play central role at the time of viral replication

Nonstructural protein	Allocated function	Reference
nsp1	IFN signaling inhibition, degradation of cellular mRNAs, translation inhibition, cell cycle arrest	Chen et al. (2020), de Groot et al. (2012), Kamitani et al. (2006), Huang et al. (2011)
nsp2	Unknown; associates with RTCs	Chen et al. (2020), de Groot et al. (2012)
nsp3	Papain-like proteinase; polyprotein processing; ADP-ribose-phosphatase (macrodomain); RNA-binding; antagonist to interferon (IFN); host innate immune response blocking	Chen et al. (2020), de Groot et al. (2012)
nsp4	Unknown; double membrane vesicle (DMV) formation	Chen et al. (2020), de Groot et al. (2012)
nsp5	Main proteinase M, polyprotein processing; inhibition of IFN signaling	Chen et al. (2020), de Groot et al. (2012)
nsp6	Unknown; DMV formation; restriction of autophagosome expansion	Chen et al. (2020), de Groot et al. (2012)
nsp7	ssRNA binding, cofactor with nsp8 and nsp12	Chen et al. (2020), de Groot et al. (2012)
nsp8	Cofactor with nsp7 and nsp12, primase	Chen et al. (2020), de Groot et al. (2012)
nsp9	ssRNA binding; associates with replication-transcription complexes (RTCs), dimerization	Chen et al. (2020), de Groot et al. (2012)
nsp10	Dodecameric zinc finger protein; associates with RTCs, stimulates nsp16 methyltransferase activity, scaffold protein for nsp14 and nsp16	Chen et al. (2020), de Groot et al. (2012)
nsp11	Unknown	Chen et al. (2020), de Groot et al. (2012)
nsp12	Primer-dependent RdRp	Chen et al. (2020), de Groot et al. (2012)
nsp13	Helicase; RNA 5′-triphosphatase	Chen et al. (2020), de Groot et al. (2012)
nsp14	$3' \rightarrow 5'$exoribonuclease; guanine-N7-methyltransferase (RNA cap formation)	Chen et al. (2020), de Groot et al. (2012)
nsp15	Endoribonuclease; evasion of dsRNA sensors	Chen et al. (2020), de Groot et al. (2012)
nsp16	Ribose-2′-*O*-methyltransferase (RNA cap formation)	Chen et al. (2020), de Groot et al. (2012)

unique enzymes implicated in one or more important step(s) in viral replication. Other nsps appear to be entirely involved in virus–host interactions (De Groot et al. 2012).

Coronavirus SARS-CoV-2 is positive ssRNA and enveloped virus. Genome encodes 16 nonstructural proteins (NSPs) and 4 essential structural proteins and several accessory proteins.

13.3 Alteration in Glycosylation Pattern of Spike Glycoproteins

The comparison of the glycosylation sites between COVID-19 and SARS-CoV of the spike glycoproteins unveiled the presence of novel glycosylation sites in COVID-19, such as NGTK, NFTI, NLTT, and NTSN (Kumar et al. 2020). The presence of novel glycosylation sites in COVID-19 might be due to the variation in sequence. Along with the deviation in reported four glycosylation sites, i.e., NGTK, NFTI, NLTT, and NTSN, COVID-19 spike glycoprotein exhibits the similar glycosylation sites, also present in SARS-CoV (NITN, NGTI, NITN, NFSQ, NESL, NCTF, and NNTV). There is a possibility of interaction between COVID-19 and host receptor through novel glycosylation sites which may have an effect on the process of internalization and linked pathogenesis (Kumar et al. 2020).

13.4 Antigenic Variation in Spike Glycoproteins

A study carried out to determine the antigenicity by comparing the antigenic variations in both spike glycoproteins of COVID-19 and SARS-CoV revealed the individuality of most of the CTL epitopes in COVID-19 compared to the SARS-CoV. However, six identical epitopes RISNCVADY, CVADYSVLY, RSFIEDLLF, RVDFCGKGY, MTSCCSCLK, and LKGVKLHY are present in spike glycoprotein of COVID-19 and SARS-CoV (Kumar et al. 2020). In COVID-19, some epitopes are identified with difference in single amino acid change. As per available information and research data of antigenicity, COVID-19 shows little antigenic similarities with SARS coronavirus. It is possible that antigenic similarity may cause similar antigenic response, and therefore, it can be used as one of the precautionary approach. Variation and similarity in spike glycoprotein and epitopes may be useful to design newer and effective vaccines (Kumar et al. 2020).

13.5 Structural Difference in Spike Glycoproteins

The length of encoded proteins in COVID-19 and SARS-like coronaviruses was found nearly same (Guo et al. 2020). However, longer spike protein of COVID-19 showed a notable difference compared to the bat SARS-like coronaviruses and SARS-CoV. On the whole 12.5% of difference in sequences of S glycoprotein and difference in minimal RBD with 23.6% may cause structural differences in spike glycoproteins of COVID-19 and SARS-CoV (Kumar et al. 2020). The studies performed to measure the average distance between the molecules of superimposed proteins, i.e., RMSD, depict the 1.38 Å local RMSD value between two glycoproteins. This illustrates regardless of 12.8% sequence variation that there is an insignificant structural divergence among the spike glycoproteins of SARS-CoV and COVID-19. The absence of structural divergence in spike glycoproteins of SARS-CoV and COVID-19 raises a hope for the treatment of brutal COVID-19. As there is structural similarity in SARS-CoV and COVID-19, the previously trialed and tested attachment inhibitor used for SARS-CoV can be used as the present choice of treatment for COVID-19 (Kumar et al. 2020).

13.6 RNA Recombination in Positive-Strand RNA Viruses

Genetic recombination is an important evolutionary mechanism in positive-strand RNA viruses. Recombination drives toward the diversity in viral genome by the creation of novel chimeric genomes (Bentley and Evans 2018; Loriere and Holmes 2012). Irrespective of single- or multiple-segment genome, RNA recombination may occur in the entire RNA viruses (Loriere and Holmes 2012). Cross-species transmission in RNA viruses is the most common approach to get entered into a new host. The process of recombination assists the entry of viruses as it facilitates viruses to explore a larger proportion of the sequence area, than mutation. This increases the probability of finding a genetic configuration that assists host adaptation (Loriere and Holmes 2012). Both replicative and non-replicative recombination mechanism works for the viruses. In replicative recombination, the major role is played by viral polymerase; however, other viral or cellular proteins may exist. On the contrary, cellular components are solely responsible for non-replicative recombination (Bentley and Evans 2018).

Recently emerged many human diseases are originated from RNA viruses. There are three mechanisms by which viruses go through evolutionary changes. These are mutations also known as antigenic drift, re-assortment (antigenic shift), and recombination (Shao et al. 2017). The emergence of viral diseases in human exhibits active viral genome recombination or re-assortment. Coronavirus that emerged in turkey was a recombinant infectious bronchitis virus which attained a spike protein-encoding gene from coronavirus 122 (Loriere and Holmes 2012). There is a possibility of emergence of COVID-19 through RNA recombination.

Executive Summary

- The third epidemic of the twenty-first century:
  - SARS-CoV was reported as the first epidemic of the twenty-first century in 2003 following by MERS-CoV in 2012.
  - COVID-19 caused by SARS-CoV-2 is the third epidemic which is declared pandemic by the WHO.
- Severe acute respiratory syndrome (SARS) and coronavirus disease 2019 (COVID-19) portray a number of similarities:
  - Virus homology: As per sequence alignment data, the spike glycoprotein sequence of COVID-19 and SARS-CoV shows 87.2% resemblance; sequences of minimal receptor-binding domain (RBD) show 83.9% similarity.
  - Presence of similar glycosylation sites and similar epitopes.
  - Routes of transmission of both the viruses (i.e., contact, droplets, and fomite).
- Along with the similarities, severe acute respiratory syndrome (SARS) and coronavirus disease 2019 (COVID-19) depict a number of dissimilarities also:
  - Such as 12.5% of difference in S glycoprotein sequence and difference in minimal RBD with 23.6%.
  - Other behavioral differences, such as rate of transmission and severity of COVID-19 is much higher than SARS-CoV.
- The emergence of novel coronavirus 2019 (COVID-19) might be due to RNA recombination as the positive-sense RNA viruses are well reported for their ability of RNA recombination which make them potent to cross the species barrier.
- The structural similarity of COVID-19 with SARS-CoV could be encashed in terms of:
  - Use of available treatment for the inhibition of attachment with the host cell. This could be an effective way to control the spread of disease.
  - Can use some drug/compound to block the entry into host cell by binding to the receptor.

13.7 Conclusions

The newly emerged coronavirus SARS-CoV-2 is strongly related to the predecessor SARS-CoV. SARS-CoV is becoming pandemic which is declared as public health emergency of global concern. The presence of novel glycosylation sites in SARS-CoV-2 makes it possible to become pandemic due to its antigenic divergence. At this time, it is important to understand and educate the mass to follow the instructions given by the authorities and avoid any social gathering. By this at least we can delay the transmission in widespread community. Slowdown of transmission rate will provide time to the researchers and scientists to prepare well and develop vaccine and treatment for this novel coronavirus. The similarity of SARS-CoV-2 to SARS-CoV in terms of antigenic sites proposes the scope of SARS-linked peptide-based vaccine to prevent COVID-19. The structural similarity of SARS-CoV-2 to SARS-CoV suggests the use of attachment inhibitor, specific to coronavirus as the current option of the treatment. Although the mechanism of species to species transfer of the SARS coronavirus is difficult to understand, SARS-CoV 2003 epidemic makes it clear that animal coronaviruses are budding threats to the human community. Still researches are going on COVID-19. Limited information on novel coronavirus makes some boundaries to explain the complete antigenicity and structure of COVID-19.

13.8 Future Perspectives

Complete genome analysis is required to understand the similarities and differences of SARS-CoV-2 with the previously reported viruses. Complete genome analysis also helps in the development of vaccines and medicines against COVID-19. It is required to understand and do the complete research on the emergence and reemergence of coronaviruses and to understand the change in proteins and genome. Transmission through crossing the species barrier is another area to focus on.

References

Adnan Shereen M, Khan S, Kazmi A, Bashir N, Siddique R (2020) COVID-19 infection: origin, transmission, and characteristics of human coronaviruses. J Adv Res 24:91–98. https://doi.org/10.1016/j.jare.2020.03.005

Banerjee A, Kulcsar K, Mishra V, Frieman M, Mossman K (2019) Bats and coronaviruses. Viruses 11:41

Belouzard S, Chu VC, Whittaker GR (2009) Activation of the SARS coronavirus spike protein via sequential proteolytic cleavage at two distinct sites. Proc Natl Acad Sci U S A 106 (14):5871–5876

Bentley K, Evans DJ (2018) Mechanisms and consequences of positive-strand RNA virus recombination. J Gen Virol 99:1345–1356

Cascella M, Ranik M, Cuomo A, Dulebohn SC, Napoli RD (2020) Features, evaluation and treatment coronavirus (COVID-19). NCBI Bookshelf: https://www.ncbi.nlm.nih.gov/books/NBK554776/

Chen Y, Liu Q, Guo D (2020) Emerging coronaviruses: genome structure, replication, and pathogenesis. J Med Virol 92:418–423

Clarivate Analytics (2020) Disease briefing: coronaviruses. https://clarivate.com/wp-content/uploads/dlm_uploads/2020/01/CORONAVIRUS-REPORT-1.30.2020.pdf

Coutard B, Valle C, Lamballerie D, Canard B, Seidah NG, DEcroly E (2020) The spike glycoprotein of the new coronavirus 2019-nCoV contains a furin like cleavage site absent in CoV of the same clade. Antivir Res 176:104742

Cui J, Li F, Shi ZL (2019) Origin and evolution of pathogenic coronaviruses. Nat Rev Microbiol 17:181–192

De Groot RJ, Baker SC, Baric R, Enjuanes L et al (2012) Family—Coronaviridae. Virus taxonomy ninth report of the International Committee on Taxonomy of Viruses, pp 806–828

Du L, He Y, Zhou Y, Liu S, Zheng BJ, Jiang S (2009) The spike protein of SARS-CoV—a target for vaccine and therapeutic development. Nat Rev Microbiol 7(3):226–236. https://doi.org/10.1038/nrmicro2090

Guo YR, Cao QD, Hong ZS, Tan YY et al (2020) The origin, transmission and clinical therapies on coronavirus disease 2019 (COVID-19) outbreak – an update on the status. Mil Med Res 7:11

Huang C, Lokugamage KG, Rozovics JM, Narayanan K, Semler BL et al (2011) SARS coronavirus nsp1 protein induces template-dependent endonucleolytic cleavage of mRNAs: viral mRNAs are resistant to nsp1-induced RNA cleavage. PLoS Pathog 7(12):e1002433. https://doi.org/10.1371/journal.ppat.1002433

Hulswit RJG, Langa Y, Bakkersa MJG et al (2019) Human coronaviruses OC43 and HKU1 bind to 9-oacetylated sialic acids via a conserved receptor-binding site in spike protein domain A. Proc Natl Acad Sci U S A 116(7):2681–2690

Human Coronavirus Types (2020) Centre for Disease control and prevention. https://www.cdc.gov/coronavirus/types.html

Kahn JS (2005) History and recent advances in coronavirus discovery. Pediatr Infect Dis J 24: S223–S227

Kamitani W, Narayanan K, Huang C, Lokugamage K et al (2006) Severe acute respiratory syndrome coronavirus nsp1 protein suppresses host gene expression by promoting host mRNA degradation. Proc Natl Acad Sci U S A 103(34):12885–12890

Kampf G, Todt D, Pfaender S, Steinmann E (2020) Persistence of coronaviruses on inanimate surfaces and their inactivation with biocidal agents. J Hosp Infect 104:246–251

Kumar S, Maurya VK, Prasad AK, Bhatt MLB, Saxena SK (2020) Structural, glycosylation and antigenic variation between 2019 novel coronavirus (2019-nCoV) and SARS coronavirus (SARS-CoV). Virus Dis. https://doi.org/10.1007/s13337-020-00571-5

Lau SKP, Chan JFW (2015) Coronaviruses: emerging and re-emerging pathogens in humans and animals. Virol J 12:209. https://doi.org/10.1186/s12985-015-0432-z

Li X, Geng M, Peng Y, Meng L, Lu S (2020) Molecular immune pathogenesis and diagnosis of COVID-19. https://doi.org/10.1016/j.jpha.2020.03.001

Lim YX, Ng YL, Tam JP, Liu DX (2016) Human coronaviruses: a review of virus–host interactions. Diseases 4:26. https://doi.org/10.3390/diseases4030026

Liu C, Zhou Q, Li Y, Garner LV, Watkins SP, Carter LJ, Smoot J, Gregg AC, Daniels AD, Jervey S, Albaiu D (2020) Research and development on therapeutic agents and vaccines for COVID-19 and related human coronavirus diseases. ACS Cent Sci 6(3):315–331. https://doi.org/10.1021/acscentsci.0c00272

Loriere ES, Holmes EC (2012) Why do RNA viruses recombine. Nat Rev Microbiol 9(8):617–626. https://doi.org/10.1038/nrmicro2614

Lu R, Zhao X, Li J, Niu P, Yang B et al (2020) Genomic characterisation and epidemiology of 2019 novel coronavirus: implications for virus origins and receptor binding. Lancet 395:565–574

Narayanan K, Ramirez SI, Lokugamage KG, Makino S (2014) Coronavirus nonstructural protein 1: common and distinct functions in the regulation of host and viral gene expression. Virus Res 202:89–100. https://doi.org/10.1016/j.virusres.2014.11.019

National Foundation for Infectious Diseases (2020) Coronavirus. https://www.nfid.org/infectious-diseases/coronaviruses/

Parrish CR, Holmes EC, Morens DM, Park EC et al (2008) Cross-species virus transmission and the emergence of new epidemic diseases. Microbiol Mol Biol Rev 72(3):457–470

Payne S (2017) Chapter 17 Coronaviridae. Viruses:149–158. https://doi.org/10.1016/B978-0-12-803109-4.00017-9

Phan MVT, Tri TN, Anh PH, Baker S et al (2018) Identification and characterization of Coronaviridae genomes from Vietnamese bats and rats based on conserved protein domains. Virus Evol 4(2). https://doi.org/10.1093/ve/vey035

Pradhan P, Pandey AK, Mishra A, Gupta P, et al (2020) Uncanny similarity of unique inserts in the 2019-nCoV spike protein to HIV-1 gp120 and Gag. https://doi.org/10.1101/2020.01.30.927871

Rasool SA, Fielding BC (2010) Understanding human coronavirus HCoV-NL63. Open Virol J 2010(4):76–84

Shao W, Li X, Goraya MY, Wang S, Chen JL (2017) Evolution of influenza A virus by mutation and re-assortment. Int J Mol Sci 18:1650

Shoeman D, Fielding BC (2019) Coronavirus envelope protein: current knowledge. Virol J 16:69

Tai W, He L, Zhang X, Pu J et al (2020) Characterization of the receptor-binding domain (RBD) of 2019 novel coronavirus: implication for development of RBD protein as a viral attachment inhibitor and vaccine. Cell Mol Immunol. https://www.nature.com/articles/s41423-020-0400-4

Wang LF, Shi Z, Zhang S, Field H, Daszak P, Eaton BT (2006) Review of bats and SARS. Emerg Infect Dis 12(12):1834–1840

World Health Organization (2020) Coronavirus disease (COVID-2019) situation reports. https://www.who.int/emergencies/diseases/novel-coronavirus-2019/situation-reports

Wu C, Liu Y, Yang Y, Zhang P et al (2020) Analysis of therapeutic targets for SARS-CoV-2 and discovery of potential drugs by computational methods. Acta Pharm Sin B. https://doi.org/10.1016/j.apsb.2020.02.008

Chapter 14
Preparing for the Perpetual Challenges of Pandemics of Coronavirus Infections with Special Focus on SARS-CoV-2

Sonam Chawla and Shailendra K. Saxena

Abstract COVID-19, arising from novel, zoonotic coronavirus-2, has gripped the world in a pandemic. The present chapter discusses the current internationally implemented pandemic preparedness strategies succeeding/recommended to curb the COVID-19 threat to humankind. The updated phase-wise categorization of a pandemic as recommended by the WHO is described, and associated innovations in surveillance, response, and medical measures/advisory in practice across the globe are elaborated. From a bird's eye view, the COVID-19 pandemic management relies on revolutionizing the disease surveillance by incorporating artificial intelligence and data analytics, boosting the response strategies—extensive testing, case isolation, contact tracing, and social distancing—and promoting awareness and access to pharmaceutical and non-pharmaceutical interventions, which are discussed in the present chapter. We also preview the economic bearing of the COVID-19 pandemic.

Keywords COVID-19 · Pandemic preparedness · Surveillance · Artificial intelligence · Infectious disease modeling · Social distancing

Abbreviations

ACE2	Angiotensin-converting enzyme 2
AI	Artificial intelligence
CoV-2	Coronavirus-2

Sonam Chawla and Shailendra K. Saxena contributed equally as first author.

S. Chawla (✉)
Department of Biotechnology, Jaypee Institute of Engineering and Technology University, Noida, India
e-mail: sonam.chawla@jiit.ac.in

S. K. Saxena
Centre for Advanced Research (CFAR), Faculty of Medicine, King George's Medical University (KGMU), Lucknow, India
e-mail: shailen@kgmcindia.edu

S. K. Saxena (ed.), *Coronavirus Disease 2019 (COVID-19)*, Medical Virology: from Pathogenesis to Disease Control, https://doi.org/10.1007/978-981-15-4814-7_14

COVID-19	Corona viral disease-2019
CPRP	COVID-19 Country Preparedness and Response Plan
CRISPR	Clustered regularly interspaced short palindromic repeats
MERS	Middle East respiratory syndrome
SARS	Severe acute respiratory syndrome
SPRS	Strategic preparedness and response plan
USFDA	United States Food and Drug Administration
WHO	World Health Organization

14.1 Introduction

The present-day pandemic spotlight on COVID-19 (coronavirus disease-2019) was earlier placed on Zika virus, H1N1, severe acute respiratory syndrome (SARS), chikangunya, Middle East respiratory syndrome (MERS), and Ebola. The "advancements" of the human race—increased urbanization, global travel, changes in land use, and fervent exploitation of the nature—are also the prime reasons for zoonosis and emergence of novel infectious diseases such as above (Madhav et al. 2017; Ahmed et al. 2019). This rapid emergence of novel infectious diseases transmitting from surrounding animal life to humans and then from human to human, traveling quickly across the globe can trigger worldwide public health emergency situations, as prevalent today (https://www.who.int/dg/speeches/detail/who-director-general-s-opening-remarks-at-the-media-briefing-on-covid-19-11March2020, http://www.emro.who.int/fr/about-who/rc61/zoonotic-diseases.html). The World Health Organization (WHO) declared COVID-19 a pandemic on March 11, 2020. MeSH database defines pandemics as—"Epidemics of infectious disease that have spread to many countries, often more than one continent, and usually affecting a large number of people." Such emergencies compromise human health, society, economics, and politics—a case in point: the COVID-19 pandemic is forecasted to cost the global economy one trillion US dollars (https://www.ncbi.nlm.nih.gov/mesh/?term=pandemics, https://news.un.org/en/story/2020/03/1059011).

As against the earlier guidelines of WHO where it classified a pandemic into six stages, the 2009 revision in pandemic descriptors and stages stands today as follows:

- *Predominantly animal infections, few human infections*. This corresponds with the stages 1–3 of earlier classification, starting with phase 1 where the virus is in its animal host and has caused no known infection in humans, phase 2 where zoonosis has occurred and the virus has caused infection in humans, and phase 3 where sporadic cases or clusters of infectious disease occur in humans. Human-to-human transmission is limited in time and space and is insufficient to cause community-level outbreaks.
- *Sustained human–human transmission*. Corresponds with the stage 4 of the classical description wherein animal–human and human–human transmissions have sustained a community-level outbreak. The risk for pandemic is greatly increased.

- *Widespread human infection* or the stage 5–6 from the classical description where the same identified virus has caused a community-level outbreak in another country in another WHO region.
- *Post-peak period* where there exists a possibility of recurrence of infection.
- *Post-pandemic phase* when the disease activity is seasonal (https://web.archive.org/web/20110910112007/http://www.who.int/csr/disease/influenza/GIPA3AideMemoire.pdf, https://www.reuters.com/article/uk-china-health-who-idUKKCN20I0PD).

At the time of writing this chapter, coronavirus-2 (CoV-2)/COVID-19, though originated in Wuhan, China, the first case being reported in November 2019, had pervaded Africa, Americas, Europe, South-East Asia, Eastern Mediterranean, and the Western Pacific nations with 191,127 confirmed cases of COVID-19 and claimed 7807 lives, globally (https://www.who.int/docs/default-source/coronaviruse/situation-reports/20200318-sitrep-58-covid-19.pdf?sfvrsn=20876712_2). Europe was declared the new epicenter of the pandemic on March 13, 2020. The number of new cases in China though declining and is believed to be in post-peak stage, the numbers are alarmingly increasing worldwide (https://www.nbcnews.com/health/health-news/europe-now-epicenter-pandemic-who-says-n1158341).

WHO and other leading epidemiology organizations unanimously agree on the indispensable role of pandemic preparation and planning at global and national levels to mitigate through the present public health emergency of COVID-19 and any future outbreaks. Pandemic preparation is not a job of single individual or organization. It requires inputs from each person susceptible to the infection agent as well as policy makers at national and international levels, frontline healthcare providers, infrastructure developers and maintenance personnel, pharmaceutical industry and researcher community, and so forth. Moreover, the pandemic preparedness plan needs constant reviewing and improvisation (https://www.ecdc.europa.eu/en/seasonal-influenza/preparedness/why-pandemic-preparedness). In line with the magnitude of the COVID-19 pandemic, worldwide action plans have been activated on national and international levels. The United Nations' *Strategic Preparedness and Response Plan (SPRS) Against COVID-19*, in layman terms, is designed to control human–human transmission, preventing outbreaks and delaying spread; provide optimal care for all patients; and minimize the impact on healthcare systems and socioeconomic activities. Under SPRS each nation is assessed for risk and vulnerability, and the resource requirements to support the country to prepare for and respond to COVID-19 are estimated. Several nations are well placed to implement this action plan with minimal support. However, otherwise partners are to be introduced to facilitate implementation of measures where there is a gap in capacity, on either a national or a subnational level, in additional support to national governments. Thus, an extensive analysis and identification of an affected nation's gaps and needs shall be the basis to develop a *COVID-19 Country Preparedness and Response Plan (CPRP)*. These CPRPs will need constant monitoring and reviewing using indicators charted in the SPRP and updated as the situation evolves (https://www.who.int/docs/default-source/coronaviruse/covid-19-sprp-unct-guidelines.pdf).

Grossly, the extent of success of each pandemic action plan stands on the following pillars:

- Surveillance of coronavirus-2 and COVID-19 infection: characterization of the virus, infection modes, diagnosing and detecting infection, contact tracing, annotation of data from confirmed cases, predicting mass infection outbreak, keeping a count, and estimation of mortality.
- Response management: bulk production and supply of protective/preventive pharmaceutical interventions or non-pharmaceutical interventions.
- Facilitating timely medical help: access to hospitals/healthcare providers, personal and public hygiene, disinfection, and quarantine services.
- Lesson learning from the present outbreak of COVID-19 to facilitate future action plans and preparedness.

Hereafter we discuss the present salient strategies under the aegis of the COVID-19 pandemic preparation plan, globally, which are helping the humankind mitigate through this emergency. We also discuss the impact of COVID-19 on world economy and its bearing on future preparedness plans.

14.2 Strategies for Pandemic Preparedness for Coronavirus

14.2.1 Updated Surveillance Systems

As defined by the WHO, surveillance during pandemics is defined as "the ongoing collection, interpretation and dissemination of data to enable the development and implementation of evidence-based interventions during a pandemic event" (https://www.who.int/influenza/preparedness/pandemic/WHO_Guidance_for_surveillance_during_an_influenza_pandemic_082017.pdf). The present-day key worldwide surveillance activities against COVID-19 include:

(a) Detection of coronavirus-2 and verification of COVID-19
(b) Risk and severity assessment
(c) Monitoring the pandemic

The rapidly expanding array of PCR/reverse transcriptase PCR-based diagnostics which are quick and efficient in identifying the virus (Pang et al. 2020; Lake 2020, https://www.finddx.org/covid-19/) are basic requirements for surveillance at every phase of the pandemic for identifying and segregating the infected from non-infected and risk assessment, as well as monitoring recurrences and seasonal disease activity. COVID-19 diagnostics have been discussed previously in this publication.

An emerging exciting field of disease surveillance is infectious disease modeling and incorporation of artificial intelligence. Notably, these are *predictive techniques* applicable to each of the three facets of disease surveillance (Siettos and Russo 2013). A classical epidemiological surveillance parameter is quantitation of R_0 (R nought, basic reproductive number) using mathematical models (https://

wwwnc.cdc.gov/eid/article/25/1/17-1901_article). R_0 is a crucial metric indicating that on average the number of new infection cases are generated by a confirmed infection case, i.e., the potential transmissibility of an infectious disease. R_0 of COVID-19 infection is estimated as 2–3.5 in the early phase, as even the asymptomatic patients or with mild pneumonia extruded large amounts of virus (Wang et al. 2020). Important characteristics of R_0 are:

- It is a dynamic number and changes with
  - Each stage of the disease
  - With interventions, e.g., vaccination, antivirals
  - Precautionary measures such as personal/community disinfection, social distancing, and travel restrictions
- With the knowledge of R_0, one can predict:
  - New cases expected on a daily basis, and hence facilitate arrangement of healthcare services and interventions locally
  - Outbreak size and the dates of peak infection of the pandemic
  - Probable decline timeline
  - The extent of vaccination coverage required to prevent future outbreaks
- An $R_0 > 1$ indicates that each infected individual is transmitting the disease to more than 1 new individual, and the infection is spreading increasingly. $R_0 = 1$ indicates stable transmission and $R_0 < 1$ indicates the decline in disease transmission (Fig. 14.1). The aim of pandemic action plans is to monitor and depreciate the $R_{0.}$

For instance, the median daily R_0 in Wuhan declined from 2.35 a week before the introduction of travel restrictions (January 23, 2020) to 1.05 one week later. The study used a stochastic transmission model of CoV-2 transmission with four datasets from within and outside Wuhan and estimated how transmission in Wuhan varied between December 2019 and February 2020 (Kucharski et al. 2020).

The most prominent mathematical model of COVID-19 infection is the *SEIR (Susceptible-Exposed-Infected-Recovered) model* put forth by Wu and coworkers and also endorsed by the WHO. This model estimates the size of epidemic in Wuhan between December 2019 and January 2020 and forecasts the extent of domestic and global public health risks of taking into account for social and non-pharmaceutical prevention interventions. It is a compartmental model comprising four compartments and the individuals comprising the sample population move through each compartment—"Susceptible" (not immune to infection) and get infected from other infection individuals and move to the "Exposed" compartment for the incubation period. Hereafter the infectious individuals move to the "Infected" compartment and eventually to the "Recovered" compartment after the disease has run its course, and they now have some immunity (Fig. 14.2). The changes in the population in each compartment are estimated using ordinary differential equations to simulate the progression of an infectious disease. The critical parameters associated with this model are:

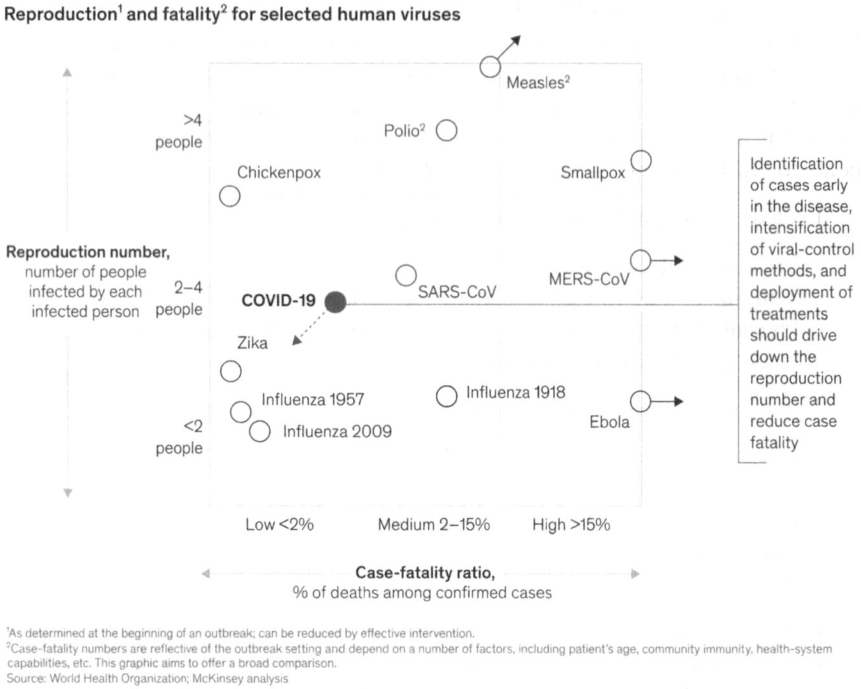

Fig. 14.1 Reproduction number (R_0) and fatality for selected human viruses. (Source: WHO)

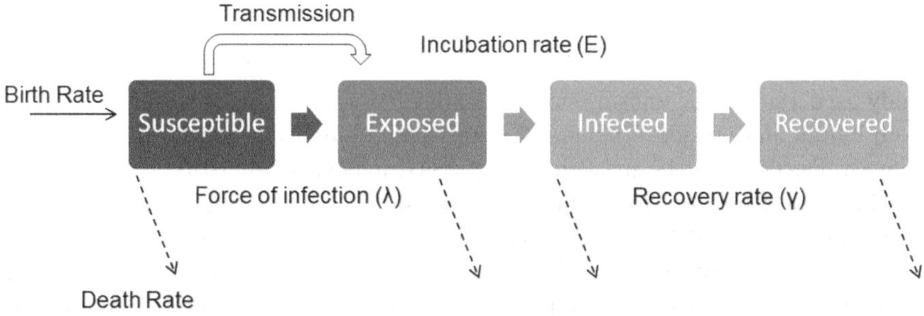

Fig. 14.2 The SEIR (Susceptible-Exposed-Infected-Recovered) mathematical model of COVID-19 infection

- Force of infection (λ) is the rate at which susceptible individuals are exposed. It depends on the transmission rate (β).
- Incubation rate (E) is the rate at which exposed people become infectious.
- Recovery rate (γ) is the rate at which infected individuals recover from the infection.

Through this model, it was predicted that the R_0 was 2.68, each confirmed case infected 2–3 other people, and the epidemic doubling time was 6.4 days. Also, the size of the outbreak in Wuhan was estimated to be up to 75,815 people (statistical uncertainty presented at 95% credible intervals). Most striking feature of this model was that it took into account the travel data from and to Wuhan over the period of study. Thus, it was able to predict that multiple major Chinese cities—Guangzhou, Beijing, Shanghai, and Shenzhen—had already imported the infection to trigger local epidemics. It also recommended that controlling the transmissibility by 25–50% could eventually rein the local epidemics, and a control of $\geq$63% would phase out the epidemics (Wu et al. 2020).

The application of artificial intelligence (AI) in close conjunction with technology in pandemic surveillance has demonstrated manifold advantages in surveillance activities as exemplified by China and other severely hit nations during the present COVID-19 outbreak. AI, data analytics, and technological support amalgamated to facilitate:

- *Track and forecast community outbreaks*:
  BlueDot is a Canadian AI company using natural language processing and machine learning algorithms to monitor news outlets, worldwide official healthcare reports in several different languages, and air-travel data and flag the mention of contagious or novel diseases such as coronavirus. Importantly this is followed by scrutiny by epidemiologists and thus also has a component of human analytics. BlueDot alerted its clients to the potential outbreak in Wuhan, China, on December 30, 2019, 9 days prior to the WHO recognized it as an epidemic (https://bluedot.global/products/).
- *Disease diagnosis and verification of infected individuals*:
  Chest computed tomography (CT) scans have been endorsed as a primary diagnostic tool for COVID-19 (Ai et al. 2020). Ali-baba group's research academy has developed a deep-learning AI-enabled system that can diagnose COVID-19 in 20 s with 96% accuracy and hence possibly automate the diagnosis activity in the face of overburdened healthcare systems. The AI system identifies an infectious individual based on the chest CT scans. The algorithm has been trained with data and CT scans from nearly 5000 confirmed coronavirus cases from across China. It can be used to track the efficacy of treatment during the course of infection as well as rapidly diagnose COVID-19 (https://www.alizila.com/how-damo-academys-ai-system-detects-coronavirus-cases/).
- *Implementing public hygiene guidelines*:
  Risk communication in places of potential communication is critical to alert the public and implement mass hygiene measures such as use of sanitizers and face masks when in public places. Google Trends was used in Taiwan to monitor the public risk awareness following the first imported case of COVID-19 which correlated with the increased search keywords "COVID-19" and "face masks." Moreover, search for "handwashing" increased coinciding with the face mask shortage. High to moderate correlations between Google relative search volume and COVID-19 cases were evident in several major cities of Taiwan (Husnayain et al. 2020). Similarly, in China an AI-based company SenseTime has developed a "Smart AI Epidemic Prevention Solutions"—a quick and effective system

based on facial recognition and thermal imaging to screen for individuals with fever in a crowd without physical contact and hence preventing transmission. It can also monitor if any individual is wearing a face mask or violating the quarantine rules (https://www.sensetime.com/en/news/view/id/140.html).

- *Contact tracing*:
  Health Code, a Chinese government monitoring system, for which users can sign up through Alipay or WeChat, assigns individuals a color code (red—14 days self-quarantine/yellow—7 days self-quarantine/green—free movement) based on their travel history, time spent in outbreak hotspots and exposure to potential carriers of the virus. The software can be used to check the color of an individual on entering their identity numbers.
- *Characterization of the CoV-2 and vaccine/therapy development*:
  The Genome Detective Coronavirus Typing Tool is a web-based, user-friendly software application that can identify the novel severe acute respiratory syndrome (SARS)-related coronavirus (SARS-CoV-2) sequences isolated in the original outbreak in China and later around the world. The tool accepts 2000 sequences per submission, analyzing them in approximately 1 min. This tool facilitates tracking of new viral mutations as the outbreak expands globally, which may help to accelerate the development of novel diagnostics, drugs, and vaccines to stop the COVID-19 disease (Cleemput et al. 2020).

  Google's AI platform DeepMind-based protein structure prediction tool AlphaFold has predicted and released the 3D structures of several understudied proteins of the CoV-2 as an open source. These can be useful in designing antivirals/vaccines against COVID-19 in future outbreaks (Jumper et al. 2020).

In summary, the new-age pandemic surveillance using AI, data analytics, mathematical modeling of infectious diseases, and risk prediction has significantly contributed to the management of the present COVID-19 pandemic and is laying the foundation for future improvisation of the pandemic action plans.

14.2.2 Strategies for the Implementation of Laboratory Diagnostic Services

Critical issues to be addressed while recruiting laboratory diagnostic services during the novel COVID-19 pandemic are:

- The authenticity of the diagnostic tool in light of the novelty of the coronavirus-2
- Bulk of samples to be processed
- The ease of obtaining the sample for use in the diagnostic tool

The coronavirus causing COVID-19 was first isolated from a clinical sample on January 7, 2020, and within weeks, several reliable and sensitive diagnostic tools were developed and deployed. By mid-January, the first RT-PCR assays for COVID-19 were accessible in Hubei. The viral sequences and PCR primers and probe sequences were open-sourced and uploaded to public platforms by the Centre for Disease Control, China. By February, there were ten kits for the detection of COVID-19 approved in China by the National Medical Products Association—six were RT-PCR kits, one isothermal amplification kit, one virus sequencing product, and two colloidal gold antibody detection kits (https://www.who.int/docs/default-source/coronaviruse/who-china-joint-mission-on-covid-19-final-report.pdf). As of today, when the disease has assumed pandemic proportions, the volume of diagnostic tools needs to be multiplied, and hence, the United States Food and Drug Administration (USFDA) has provided regulatory relief to several testing companies like ThermoFisher, Hologic, and LabCorp under the Emergency Use Authorizations, to facilitate ease of diagnosis. Also it is keeping a tight watch on fraudulent companies claiming to sell interventions against COVID-19 (https://www.fda.gov/emergency-preparedness-and-response/mcm-issues/coronavirus-disease-2019-covid-19). Moreover, the world's first CRISPR (clustered regularly interspaced short palindromic repeats)-based diagnostic kit has been developed for COVID-19 by Mammoth Biosciences, USA, and University of California. Notably, the diagnostic kit is a simple strip-based assay and is easy to use and allows rapid detection without the need of transporting samples over long distances. The kit is still under approval evaluation by the FDA (https://www.medrxiv.org/content/10.1101/2020.03.06.20032334v1).

Nose and mouth swabs, the most widespread samples for CoV-2 diagnostics, require trained personnel to procure samples. However, the study by To and coworkers recommends saliva as an easy-to-procure, noninvasive sample not requiring any trained personnel in COVID-19 screening. They detected CoV-2 in 11 out of 12 patient samples and also could trace the declining titers post-hospitalization (To et al. 2020).

Additionally, a long-term goal of evolving the diagnostics for this novel disease is to develop a prognostic marker of COVID-19. Although it is too soon to say, Qu and coworkers have proposed platelet counts and platelet to lymphocyte ratios as prognostic marker to distinguish between severe and non-severe patients. Severe patients had higher platelet peaking and platelet to lymphocyte ratio correlating with deranged chest CT and longer hospital stays against the lower platelet peaks and platelet to lymphocyte ratios and lesser hospitalization stay (Qu et al. 2020).

14.2.3 Management of Bulk Vaccines and Antiviral Drugs

Vaccines and antiviral drugs are the prophylactic and therapeutic measures against a viral disease CoV-2. On March 17, 2020, Pfizer Inc. and BioNTech announced to co-develop and distribute a potential mRNA-based coronavirus vaccine which is likely to enter clinical testing by the end of April 2020 (https://www.pfizer.com/news/press-release/press-release-detail/pfizer_and_biontech_to_co_develop_potential_covid_19_vaccine). However, in pandemic scenarios of new infectious diseases such as COVID-19, vaccine supplies will be limited or nonexistent at the early phase in lieu of the novelty of the disease and the unpredictability of the pandemic occurrence. Thus, vaccines cannot be stockpiled, and production can only start once the novel virus has been recognized. With the current state-of the-art, the first doses of vaccine are not likely to become available in the early months of the pandemic. However, the pandemic action plans have accounted for forward planning to increase the likelihood that the vaccine will progressively become available as the pandemic unfolds. Importantly, national or regional priorities need to be fixed in the action plan for the rational use of the building/limited supply of the novel vaccine. Also, production and use of vaccines during the inter-pandemic period will influence their availability during a pandemic. Thus, improving the infrastructure and logistics for vaccine production, administration, cold-chain, and professional training with the novel vaccines are important to avert/cruise through future outbreak events. The WHO advisory on prioritizing the population groups are as enlisted in descending order. However, the priorities need tailoring in each country/region according to local needs and epidemiological circumstances.

Recommended prioritizing of the groups is as follows:

- Healthcare workers and essential service providers
- Groups at high risk of death and severe complications requiring hospitalization
- Individuals (adults and children aged more than 6 months) in the community who have chronic cardiovascular, pulmonary, metabolic or renal disease, or are immunocompromised
- Persons without risk factors for complications (https://www.who.int/csr/resources/publications/influenza/11_29_01_A.pdf)

Antivirals are a crucial adjunct to vaccination as a potential strategy for managing COVID-19. Several drugs such as chloroquine, arbidol, remdesivir, and favipiravir are currently undergoing clinical studies to establish their efficacy and safety against COVID-19. It is important to establish a regular supply chain of antivirals and a high surge capacity in the face of CoV-2 pandemic and future outbreaks. Antivirals have a significant impact in reducing morbidity and mortality in light of unlikelihood of availability of vaccine against in early phases of pandemic. It is important to evaluate the non-interference of the antiviral interventions with the eventual vaccination, as well as the epidemiology of the group of individuals most seriously affected. It is also advisable to make available the information about the performance characteristics, side effects, and costs of antiviral therapy to public. Also the commonly used neuraminidase inhibitors in influenza pandemics are ineffective in the case of CoV-2 mediated infection and outbreaks (Pang et al. 2020; Dong et al. 2020a, https://www.who.int/csr/resources/publications/influenza/11_29_01_A.pdf). The incorporation of antiviral therapy can be categorized as a prophylactic and for treatment use. As with vaccines, prioritizing of groups for antiviral therapy is advised as follows:

- Essential service providers, including healthcare workers (prophylaxis or treatment). Especially, healthcare providers are in a position to be in direct contact with infectious individuals and are thus entitled to priority antiviral therapy. Other

community services such as those responsible for vaccine manufacture and delivery and personnel responsible for enforcing law and order and public safety.
- Groups at high risk of death and severe complications requiring hospitalization. The goal of prophylaxis or treatment here is to rein the mortality and morbidity. Thus, high-risk persons living in the community outbreaks, seriously ill hospitalized patients, patients for whom a potential CoV-2 vaccination is contraindicated are prioritized.
- Persons without known risk factors for complications from COVID-19. Here the approach is generally therapeutic and aims to rein the morbidity and rationalize the use of healthcare resources such as antibiotics. Though, the logistics of this strategy are extensive and expensive (requires large quantities of antivirals and access to healthcare service providers), it is most likely to limit the economic and social destabilization associated with a pandemic (Monto 2006; Henry 2019, https://www.who.int/csr/resources/publications/influenza/11_29_01_A.pdf).

A major step in ensuring bulk supplies of antivirals and vaccines is the setting up of *medical stockpiles* as a part of the pandemic preparedness plans. The USA has constituted a Strategic National Stockpile which is the nation's largest supply of potentially life-saving pharmaceuticals and medical supplies—antibiotics, chemical antidotes, antitoxins, vaccines, life-support medication, IV administrations, airway maintenance supplies, and other emergency medical and surgical items, for use in a public health emergency severe enough to deplete the local supplies. The facility also houses a data bank of other stockpiles and supply agencies, so that any emergency requirements can be procured in the shortest possible time (https://www.phe.gov/about/sns/Pages/default.aspx).

14.2.4 Promoting facilities for treatment and hospitalization, healthcare personnel, maintenance of hygiene and disinfection during pandemics

Response component of a pandemic action plan constitutes

- Rapid facilitation of treatment/prophylaxis, hospital centers, quarantine centers
- Caring for the patients as well as the health service providers to ensure uninterrupted care
- Communicating awareness about public and personal hygiene and implementing measure to ensure personal and public hygiene

In the face of a highly infectious disease causing novel virus, China has successfully executed an ambitious, swift, and aggressive disease containment effort in the history of mankind. The laudable of its responses to facilitate timely medical care for infected was construction of two dedicated hospitals—1000-bed Huoshenshan facility and the 1600-bed Leishenshan Hospital in 2 weeks (https://www.wsj.com/articles/how-china-can-build-a-coronavirus-hospital-in-10-days-11580397751). Moreover, it ensured coordinated medical supplies, reserve beds were used and relevant premises were repurposed medical care facilities, and prices of commodities were controlled to ensure the smooth operation of the society (https://www.who.int/docs/default-source/coronaviruse/who-china-joint-mission-on-covid-19-final-report.pdf).

Ensuring the care of the medical service providers is a key response strategy to warrant efficient care for the general public. Under the China's response plan, the healthcare workers were facilitated with personal protective equipment. Nosocomial infections accounted were reported to be nearly 2055 from 476 hospitals across China. The majority of this nosocomial infection (88%) were reported from Hubei. A deeper contact tracing indicates infection of the healthcare worker from households than the workplace and were pinpointed to the early stages of the epidemic when understanding of the COVID-19 transmission and medical supplies were limited (https://www.who.int/docs/default-source/coronaviruse/who-china-joint-mission-on-covid-19-final-report.pdf). In fact a novel infection control system for averting nosocomial infections of COVID-19 was proposed by Chen and coworkers, titled "the observing system." Designated personnel called "infection control observers" were appointed by the Department of Infection Control and Nursing in Guangdong Second Provincial General Hospital who underwent training to familiarize infection control requirements in the negative pressure isolation wards. The wards were under camera surveillance and the infection control observer monitors medical staff in real-time via computer monitors outside the ward. The observer ensures normal operation of the negative pressure isolation wards, supervise the implementation of disinfection, ensure a sufficient supply of protective materials, arrange specimens for inspection, and relieve anxiety of the medical personnel while treating patients (Chen et al. 2020).

Personal and public hygiene/disinfection implementation is a key step to control transmission and prevent community outbreaks. Good hand hygiene and respiratory hygiene have been aggressively promoted by WHO worldwide in this COVID-19 pandemic. The basic personal and public hygiene practices are depicted in Fig. 14.3.

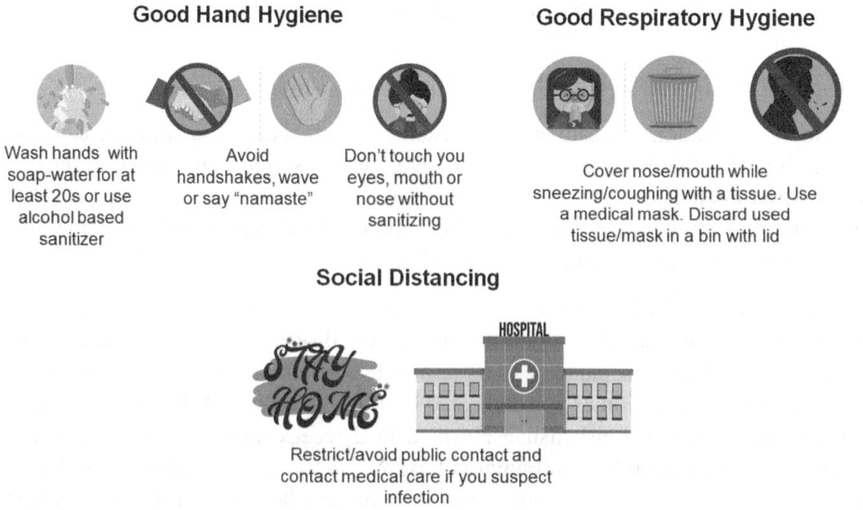

Fig. 14.3 The basic personal and public hygiene practices

A key parameter to consider while disinfecting in households and public places hence avoid contact with infected surfaces is the estimation of decay rates of CoV-2 in aerosols and on various materials composing the surfaces. It has been estimated that CoV-2 in an aerosol (<5 μm, similar to those observed in samples obtained from the upper and lower respiratory tract in humans) was viable for up to 3 h, up to 4 h on copper surfaces, up to 24 h on cardboard, and 2–3 days on plastic and stainless steel (van Doremalen et al. 2020). On the basis of these findings, disinfection protocols can be set up for public places/medical facilities depending on the surfaces involved, and even in households' frequently touched surfaces like door handles, slabs, and tables be disinfected.

14.3 Global Management Strategies for Coronavirus Infection

The exponential transmission of CoV-2—starting with few infectious individuals with COVID-19 quickly increasing manifold in a geographical location within a short time—is a recurring observation (Dong et al. 2020b, https://www.who.int/emergencies/diseases/novel-coronavirus-2019/situation-reports, https://www.ecdc.europa.eu/en/publications-data/download-todays-data-geographic-distribution-covid-19-cases-worldwide).

Figure 14.4 depicts the exponential surge in new confirmed cases per day plotted against time lapse post outbreak. The highly implemented global COVID-19 management strategies of social distancing, travel restrictions, implementation of personal and public hygiene (non-pharmaceutical interventions), and pursuit of pharmaceutical interventions aim to delay the peaking of the outbreak, avoid burden on the healthcare infrastructure and personnel to ensure quality care for all in need, and rein overall mortality and declined health effects. This phenomenon has been described as "flattening of the curve" and is much shared on social media platforms facilitating awareness in the public (https://www.cdc.gov/flu/pandemic-resources/pdf/community_mitigation-sm.pdf). Hereafter, we discuss the key, successful globally adopted strategies for COVID-19 management.

14.3.1 Social Distancing

It is a non-pharmaceutical infection prevention and control intervention to avoid or control contact between infectious and uninfected individuals to rein the disease transmission in a community eventually culminating into decreased infection spread, morbidity and mortality. The individuals infected with CoV-2 shed the virus from their respiratory tract during the early infection stage when there are minor clinical manifestations leading to the extensive community transmission. While practicing social distancing, each healthy individual behaves like an infected individual, self-

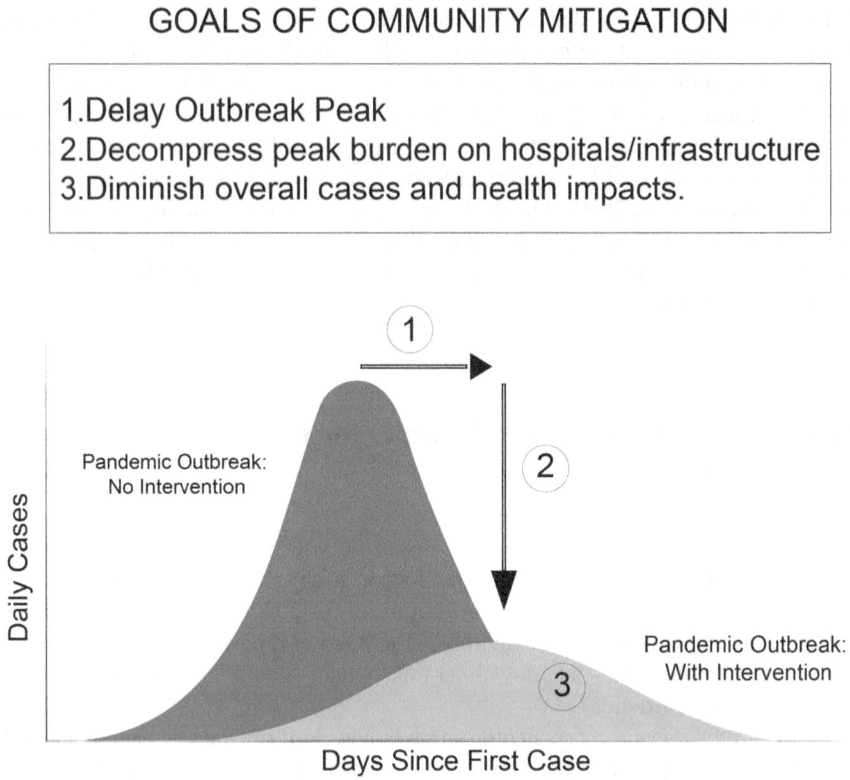

Fig. 14.4 Exponential surge in new confirmed cases per day plotted against time lapse post outbreak

restricting contact with others. The main advantage is gleaned when an individual in incubation period/early infectious stage of COVID-19 restricts coming in contact with other healthy individuals. This strategy involves the policymakers, and strict advisories/orders are issued under the pandemic action plan to restrict public gatherings and events, shutting down of educational institutes/offices/restaurants, avoiding nonessential travel and use of public transport, maintaining a distance of 1 m at least between two individuals, restricting visit to hospitals, and avoiding online shopping. Additionally, the elderly and individuals with hypertension, cardiovascular disease, diabetes, chronic respiratory diseases, and cancer are at a higher mortality risk (https://www.mohfw.gov.in/SocialDistancingAdvisorybyMOHFW.pdf, https://onlinelibrary.wiley.com/doi/pdf/10.1111/ijcp.13501; Zou et al. 2020).

14.3.2 Extensive Testing, Case Isolation, and Contact Tracing

The WHO Director-General at a media briefing on COVID-19 on March 16, 2020, recommends *testing* of any suspected case, *isolation* to break the chain of transmission, and *contact tracing* for the prior 2 days the subject came in contact with and then testing them as well (https://www.who.int/dg/speeches/detail/who-director-general-s-opening-remarks-at-the-media-briefing-on-covid-19%2D%2D-16-march-2020). A case in point—South Korea carried out 286,716 COVID-19 tests until March 17, 2020—second highest across the world trailing behind China (https://ourworldindata.org/coronavirus-testing-source-data), combined with strict social distancing implementation and is now seeing a fall in the number of COVID-19 cases (https://ourworldindata.org/grapher/daily-cases-covid-19-who?time=1..52&country=KOR, https://www.sciencemag.org/news/2020/03/coronavirus-cases-have-dropped-sharply-south-korea-whats-secret-its-success). Social distancing combined with case isolation and contact tracing has demonstrated its effectiveness in controlling the transmission from imported infection cases to the community transmission scenario (Wilder-Smith and Freedman 2020). The relevance of case isolation and contact tracing are highlighted in a study by Hu and coworkers where they have performed a clinical characterization of 24 asymptomatic patients—confirming infection by a laboratory-confirmed positive for the COVID-19 virus nucleic acid from pharyngeal swab samples. The asymptomatic cases did not present any obvious symptoms while nucleic acid screening. About 20.8% developed classical symptoms of fever, cough, and fatigue, during hospitalization; 50.0% presented CT images of ground-glass chest and 20.8% presented stripe shadowing in the lungs; 29.2% cases presented normal CT image and had no symptoms during hospitalization, comprising the younger subset (median age: 14.0 years). None developed severe COVID-19 pneumonia, and no mortality was observed. However, epidemiological investigation indicated at typical asymptomatic transmission to the cohabiting family members, which even caused severe COVID-19 pneumonia. This study puts a spotlight on close contact tracing and longitudinally surveillance via virus nucleic acid tests. Case isolation and continuous nucleic acid tests are also recommended (Hu et al. 2020). De-isolating of suspect cases where the first confirmatory test has returned negative is also an emerging issue of concern and needs improvement in testing capacity as well as strict implementation of social distancing during pandemic scenarios (Tay et al. 2020).

14.3.3 Travel Restrictions

Approaching January 13, 2020, the daily risk of exporting at least a one COVID-19 infected individual from mainland China through international travel exceeded 95%. For containment of the global spread of CoV-2 and a COVID-19 epidemic, China implemented border control measures—airport screening and travel restrictions, and were later also adopted by several other countries. Wells and coworkers in their study in Proceedings of Natural Academy of Sciences used daily incidence data of COVID-19 outbreak from China from December 8, 2019, to February 15, 2020, as well as airline network data, to predict the number of exported cases with and without measures of travel restriction and screening. The group put forth that the lockdown of Wuhan and 15 more cities in Hubei province on January 23–24, 2020, averted export of 779 more COVID-19 cases by mid-February. However, it is to be highlighted that these travel restrictions and lockdown measures only slowed the rate of infection exportation from China to other countries. The global spread of COVID-19 was based on most cases arriving during the asymptomatic incubation period. The authors recommend rapid contact tracing at the epicenter and at importation sites to limit human-to-human transmission outside of the location of first outbreak (Wells et al. 2020).

14.3.4 Foiling Myths to Control Mass Hysteria

The COVID-19 pandemic in today's times of excessive electronic connectivity and several social media platforms has a potential to alleviate the global mental stress levels. Especially, if local myths and rumors are circulated in the already home-quarantined population, restriction in fear of the disease can lead to additional mass hysteria. Thus it is important to create awareness and dispel any negative myth, as put forth by the WHO in context of CoV-2 (https://www.who.int/emergencies/diseases/novel-coronavirus-2019/advice-for-public/myth-busters). Some myths dispelled are:

- COVID-19 virus can be transmitted in areas with hot and humid climates or cold and snowy climates. CoV-2 can be transmitted in all areas.
- Hot bath cannot prevent CoV-2 transmission.
- CoV-2 is not transmitted by mosquito bites. It is a respiratory illness and respiratory hygiene is a key protective strategy.
- Heat generated from hand dryers is not effective against killing CoV-2 on your hands. The only way to disinfect hands is washing with soap-water or alcohol-based sanitizers.
- Thermal scanners are effective in detecting only the elevated body temperatures. A confirmatory nucleic acid test is the most guaranteed test.
- All age groups can be infected by CoV-2. Elderly people and people with pre-existing medical conditions—asthma, diabetes, cardiovascular disease— are more vulnerable to exhibit severe COVID-19. Thus, WHO advises people from all age groups to protect themselves by implementing good hand hygiene and respiratory hygiene.
- SARS-CoV-2 is not a bioengineered organism arising out of manipulations from earlier SARS-CoV. Andersen and coworkers have analyzed the genome sequence

of the novel coronavirus-2 and compared it with several other zoonotic viruses. They claim that the viral–human point of contact—the high affinity receptor binding domain of CoV-2 binding with the angiotensin-converting enzyme 2 (ACE2) of the target host—is acquired through natural selection. Also they have shed light of probable animal host being *Rhinolophus affinis*, as another virus RaTG13, sampled from a bat, is ~96% identical overall to CoV-2 (https://www.nature.com/articles/s41591-020-0820-9).

14.4 Economic Consequences of Pandemic Risk

The COVID-19 pandemic can cause short-term fiscal distress and longer-term damage to the global economic growth. If we trace the economic trajectory of a pandemic:

- Early phase measures to contain/limit outbreak—shutting down of workplaces and public businesses, mobilization of healthcare supplies and surveillance activities, contact tracing, social distancing by isolation of contacts/quarantines incur significant human resource and staffing costs (Achonu et al. 2005).
- As the scale of the epidemic expands, new medical infrastructure will need to be constructed to manage building number of confirmed infected individuals, the surge in demand for medical consumables can increase the health system expenditures (Herstein et al. 2016).
- The quarantine and lockdown impositions disrupt trans-national supply chains, transportation industry, agriculture, entertainment, and travel industry. Hoarding and black marketing of essential and medical commodities are expected.
- Behavioral changes during pandemic—avoiding association with other people to avoid infection—are the major determinant for economic impact of pandemics (Fig. 14.5) and not mortality. The behavioral change fear driven which in turn is driven by awareness and ignorance (Burns et al. 2008).

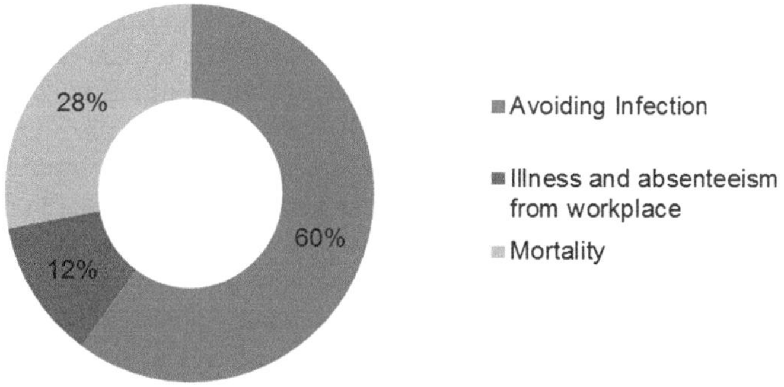

Fig. 14.5 Pandemic outcomes leading to economic impact. During pandemics, economic disturbances are an outcome of averting public contact/fear to avoid infections, absent from workplace due to illness, and mortality (https://www.worldbank.org/content/dam/Worldbank/document/HDN/Health/WDR14_bp_Pandemic_Risk_Jonas.pdf)

- National and international relief funds have been constituted to mitigate the world through the COVID-19 pandemic. For example, the SPRP estimates a funding support of nearly US$675 million for key response efforts against COVID-19 in countries most in need of help through April 2020. A first-of-its-kind COVID-19 Solidarity Response Fund was constituted for arranging for funding support through individual and organizational donations on March 13, 2020. The COVID-19 Solidarity Response Fund is hosted by two foundations, the UN Foundation (registered in the United States) and the Swiss Philanthropy Foundation (registered in Switzerland) in association with the WHO (https://www.who.int/emergencies/diseases/novel-coronavirus-2019/donate).
- Increased fiscal distress due to decrease in tax revenues and increased expenditures are expected in low middle-income countries, where fiscal framework is weak. This situation aroused earlier in West Africa Ebola epidemic in Liberia: response costs surged, economic activity slowed, and quarantines and curfews compromised government capacity to collect revenue (https://www.worldbank.org/content/dam/Worldbank/document/HDN/Health/WDR14_bp_Pandemic_Risk_Jonas.pdf).
- The high-income countries in scenario of a moderate pandemic can offset fiscal distress by providing official development assistance to affected countries and budgetary support. However, during a severe pandemic, a high-income country will also confront the same fiscal stresses and may be unwilling to provide assistance.

14.5 Conclusions and Future Perspectives

On March 19, 2020, worldwide 191,127 confirmed cases were reported, mortality was 7807 against 118,319 confirmed cases, and mortality of 280 on March 13, 2020 when the COVID-19 crisis was declared a pandemic (https://www.who.int/emergencies/diseases/novel-coronavirus-2019/situation-reports). It is increasingly becoming evident that no region of the world will remain untouched by CoV-2 invasion. However, pandemic preparedness is the key to tackle the present-day COVID-19 health emergency worldwide. The rapid and effective enforcement of existing international and national action plans, as well as parallel review and improvisation, is facilitating the affected countries to contain transmission and possibly delay the peak of outbreak and mortality and garner recovery. It is notable that in the present pandemic scenario, innovative AI-powered surveillance, quick and strategic response actions—the trinity of testing-isolation-contact tracing, committed social distancing measures—travel restrictions, self-isolation, implementation of personal and public hygiene, and extensive mobilization of medical care facilities are helping the world mitigate through. The most insightful trend emerging from the global clinical and epidemiological data is that the identification of asymptomatic infected individuals is a crucial step to contain community outbreaks. For preventing future outbreaks of CoV-2 infection, high-volume cutting-edge investigations are warranted in understanding the COVID-19 pathology, CoV-2 origin, biology, structural data of potential surface antigens, and precise anti-CoV-2 antiviral therapies. Although the global economy is suffering at the hands of CoV-2, it is important to review the current action plans and suitably improvise the future action plans to avoid potential recurrence.

Executive Summary

Pandemics are unforeseen. National and international preparedness are crucial to tackle a pandemic. World over, humanity is grappling with COVID-19. The pandemic preparedness charter prescribed by international and national agencies to tackle COVID-19 is evolving on the go. The key facets of pandemic preparedness emerging from global success and failure scenarios are:

- Active surveillance employing state-of-the art technology:
  - Development of mathematical models for simulating the infection of CoV-2 in a given country. The models can help predict the basic reproductive number which in turn facilitates monitoring the pandemic on day-to-day basis.
  - Incorporation of contactless artificial intelligence-based technologies for mass thermal screening, track and forecast community outbreaks, implement public hygiene and use of masks, and contact tracing.
- Expanding the diagnose capacity/testing for CoV-2 (the infectious agent) to catch any asymptomatic carriers, but at the same time ensuring:
  - Precision of the test.
  - Test is adaptable to processing bulk samples.
  - Easy procurement of sample from suspected infected individuals.
- Management of bulk antiviral interventions and vaccines:
  - National or regional priorities need to be fixed for rational use of antiviral/emerging vaccines
  - Prioritizing the population has been recommended: the healthcare workers and essential service providers are top priority.
  - Medical stockpiles should be established.
- Promoting facilities for healthcare/healthcare workers and facilitating public disinfection and hygiene via:
  - Rapid expansion of hospitalization facilities for treatment and isolation/ quarantine centers.
  - Ensuring minimal nosocomial infections to the healthcare workers.
  - Issuing and implementation of guidelines for ensuring good hand and respiratory hygiene.

(continued)

  - Public disinfection (and personal household as well) based on the established life of CoV-2 on different surfaces.
- Globally successful strategies are:
  - Social distancing and lockdowns to ensure minimal human–human contact.
  - Travel restrictions to facilitate containment.
  - Trifecta of test–isolate–contact tracing.
- The economic cost of the COVID-19 pandemic management is expanding due to:
  - Disruption of economic activities due to implementation of social distancing via lockdowns.
  - Battling the diseased and the increasing mortality.
  - Diversion of resources to expansion of healthcare systems.
  - International and national relief funds are being constituted.
  - Fiscal reliefs and aid from developed nations to the developing or underdeveloped countries can release the surmounting economic pressure.

References

Achonu C, Laporte A, Gardam MA (2005) The financial impact of controlling a respiratory virus outbreak in a teaching hospital: lessons learned from SARS. Can J Public Health 96(1):52–54

Ahmed S, Dávila JD, Allen A, Haklay MM, Tacoli C, Fèvre EM (2019) Does urbanization make emergence of zoonosis more likely? Evidence, myths and gaps. Environ Urban 31(2):443–460. https://doi.org/10.1177/0956247819866124

Ai T, Yang Z, Hou H, Zhan C, Chen C, Lv W, Tao Q, Sun Z, Xia L (2020) Correlation of chest CT and RT-PCR testing in coronavirus disease 2019 (COVID-19) in China: a report of 1014 cases. Radiology 26:200642. https://doi.org/10.1148/radiol.2020200642

Burns A, van der Mensbrugghe D, Timmer H (2008) Evaluating the economic consequences of avian influenza. World Bank, Washington, DC

Chen X, Tian J, Li G, Li G (2020. pii:S1473-3099(20)30110-9) Initiation of a new infection control system for the COVID-19 outbreak. Lancet Infect Dis. https://doi.org/10.1016/S1473-3099(20)30110-9

Cleemput S, Dumon W, Fonseca V, Karim WA, Giovanetti M, Alcantara LC, Deforche K, de Oliveira T (2020. pii: btaa145) Genome Detective Coronavirus Typing Tool for rapid identification and characterization of novel coronavirus genomes. Bioinformatics. https://doi.org/10.1093/bioinformatics/btaa145

Dong L, Hu S, Gao J (2020a) Discovering drugs to treat coronavirus disease 2019 (COVID-19). Drug Discov Ther 14(1):58–60. https://doi.org/10.5582/ddt.2020.01012

Dong E, Du H, Gardner L (2020b. pii: S1473-3099(20)30120-1) An interactive web-based dashboard to track COVID-19 in real time. Lancet Infect Dis. https://doi.org/10.1016/S1473-3099(20)30120-1

van Doremalen N, Bushmaker T, Morris DH, Holbrook MG, Gamble A, Williamson BN, Tamin A, Harcourt JL, Thornburg NJ, Gerber SI, Lloyd-Smith JO, de Wit E, Munster VJ (2020) Aerosol and surface stability of SARS-CoV-2 as compared with SARS-CoV-1. N Engl J Med. https://doi.org/10.1056/NEJMc2004973

Henry B (2019) Canadian pandemic influenza preparedness: antiviral strategy. Can Commun Dis Rep 45(1):38–43. https://doi.org/10.14745/ccdr.v45i01a05

Herstein JJ, Biddinger PD, Kraft CS, Saiman L, Gibbs SG, Smith PW, Hewlett AL, Lowe JJ (2016) Initial costs of Ebola treatment centers in the United States. Emerg Infect Dis 22(2):350–352. https://doi.org/10.3201/eid2202.151431

Hu Z, Song C, Xu C, Jin G, Chen Y, Xu X, Ma H, Chen W, Lin Y, Zheng Y, Wang J, Hu Z, Yi Y, Shen H (2020) Clinical characteristics of 24 asymptomatic infections with COVID-19 screened among close contacts in Nanjing, China. Sci China Life Sci. https://doi.org/10.1007/s11427-020-1661-4

Husnayain A, Fuad A, Su EC (2020. pii: S1201-9712(20)30140-5) Applications of google search trends for risk communication in infectious disease management: a case study of COVID-19 outbreak in Taiwan. Int J Infect Dis. https://doi.org/10.1016/j.ijid.2020.03.021

Jumper J, Tunyasuvunakool K, Kohli P, Hassabis D, The AlphaFold Team (2020) Computational predictions of protein structures associated with COVID-19. DeepMind website. https://deepmind.com/research/open-source/computational-predictions-of-protein-structures-associated-with-COVID-19

Kucharski AJ, Russell TW, Diamond C, Liu Y, Edmunds J, Funk S, Eggo RM. Centre for Mathematical Modelling of Infectious Diseases COVID-19 Working Group(2020. pii: S1473-3099(20)30144-4) Early dynamics of transmission and control of COVID-19: a mathematical modelling study. Lancet Infect Dis. https://doi.org/10.1016/S1473-3099(20)30144-4

Lake MA (2020) What we know so far: COVID-19 current clinical knowledge and research. Clin Med (Lond) 20(2):124–127.. pii: clinmed.2019-coron. https://doi.org/10.7861/clinmed.2019-coron

Madhav N, Oppenheim B, Gallivan M, Mulembakani P, Rubin E, Wolfe N (2017) Chapter 17: Pandemics: risks, impacts, and mitigation. In: Jamison DT, Gelband H, Horton S, Jha P, Laxminarayan R, Mock CN, Nugent R (eds) Disease control priorities: improving health and reducing poverty, 3rd edn. The International Bank for Reconstruction and Development/The World Bank, Washington, DC

Monto AS (2006 Jan) Vaccines and antiviral drugs in pandemic preparedness. Emerg Infect Dis 12 (1):55–60

Pang J, Wang MX, Ang IYH, Tan SHX, Lewis RF, Chen JI, Gutierrez RA, Gwee SXW, Chua PEY, Yang Q, Ng XY, Yap RK, Tan HY, Teo YY, Tan CC, Cook AR, Yap JC, Hsu LY (2020) Potential rapid diagnostics, vaccine and therapeutics for 2019 novel coronavirus (2019-nCoV): a systematic review. J Clin Med 9(3). pii: E623. https://doi.org/10.3390/jcm9030623

Qu R, Ling Y, Zhang YH, Wei LY, Chen X, Li X, Liu XY, Liu HM, Guo Z, Ren H, Wang Q (2020) Platelet-to-lymphocyte ratio is associated with prognosis in patients with Corona Virus Disease-19. J Med Virol. https://doi.org/10.1002/jmv.25767

Siettos CI, Russo L (2013) Mathematical modeling of infectious disease dynamics. Virulence 4 (4):295–306. https://doi.org/10.4161/viru.24041

Tay JY, Lim PL, Marimuthu K, Sadarangani SP, Ling LM, Ang BSP, Chan M, Leo YS, Vasoo S (2020. pii: ciaa179) De-isolating COVID-19 suspect cases: a continuing challenge. Clin Infect Dis. https://doi.org/10.1093/cid/ciaa179

To KK, Tsang OT, Chik-Yan Yip C, Chan KH, Wu TC, Chan JMC, Leung WS, Chik TS, Choi CY, Kandamby DH, Lung DC, Tam AR, Poon RW, Fung AY, Hung IF, Cheng VC, Chan JF, Yuen KY (2020. pii: ciaa149) Consistent detection of 2019 novel coronavirus in saliva. Clin Infect Dis. https://doi.org/10.1093/cid/ciaa149

Wang Y, Wang Y, Chen Y, Qin Q (2020) Unique epidemiological and clinical features of the emerging 2019 novel coronavirus pneumonia (COVID-19) implicate special control measures. J Med Virol. https://doi.org/10.1002/jmv.25748

Wells CR, Sah P, Moghadas SM, Pandey A, Shoukat A, Wang Y, Wang Z, Meyers LA, Singer BH, Galvani AP (2020. pii: 202002616) Impact of international travel and border control measureson the global spread of the novel 2019 coronavirus outbreak. Proc Natl Acad Sci U S A. https://doi.org/10.1073/pnas.2002616117

Wilder-Smith A, Freedman DO (2020) Isolation, quarantine, social distancing and community containment: pivotal role for old-style public health measures in the novel coronavirus (2019-nCoV) outbreak. J Travel Med 27(2). https://doi.org/10.1093/jtm/taaa020

Wu JT, Leung K, Leung GM (2020) Nowcasting and forecasting the potential domestic and international spread of the 2019-nCoV outbreak originating in Wuhan, China: a modelling study. Lancet 395(10225):689–697. https://doi.org/10.1016/S0140-6736(20)30260-9. Erratum in: Lancet. 2020 Feb 4

Zou L, Ruan F, Huang M, Liang L, Huang H, Hong Z, Yu J, Kang M, Song Y, Xia J, Guo Q, Song T, He J, Yen HL, Peiris M, Wu J (2020) SARS-CoV-2 viral load in upper respiratory specimens of infected patients. N Engl J Med. https://doi.org/10.1056/NEJMc2001737

Chapter 15
Preparing Children for Pandemics

Rakhi Saxena and Shailendra K. Saxena

Abstract COVID-19 has given us a food for thought that whether we are prepared for such pandemics or not. Developed nations may say that they have enough resources to tackle such situations, but when it comes to the physical and emotional security of the children, even they have to think manifolds because preparing children for such pandemics need lot of effort and apt planning. The purpose of this chapter is to reflect issues related to children during any infectious disease outbreak like COVID-19. Adults are mature enough to control their emotions and can act patiently, but immature minds are always perplexed and act in a very clingy way when some adverse situation is thrown to them. Children are ardent observers and act according to the reactions of the folks around them. It is difficult for them to conceal their behavior, and it is difficult for parents as well to manage their anxiety levels. During crisis period when social distancing and refrained outdoor activities have brought our children into a knotty situation, it is necessary that we understand their fears and myths, try to resolve their concerns in a polite way and strengthen their minds. At this stage the role of parents, teachers, educational institutes, social media, and international children's organizations need to be redefined. The importance of the decisions taken by the governing bodies should be explained to the children in an effective way, so that they do not panic and reflect a brave attitude. UNICEF in collaboration with international health support systems and departments has a major role to play. Appropriate planning for preparing the children for pandemics has to be incorporated in our system, so that any future crisis can be dealt in an easier way. Schools colleges and other educational institutes should

Rakhi Saxena and Shailendra K. Saxena contributed equally as first author.

R. Saxena (✉)
City Montessori School, Lucknow, India

S. K. Saxena
Centre for Advanced Research (CFAR), Faculty of Medicine, King George's Medical University (KGMU), Lucknow, India
e-mail: shailen@kgmcindia.edu

S. K. Saxena (ed.), *Coronavirus Disease 2019 (COVID-19)*, Medical Virology: from Pathogenesis to Disease Control, https://doi.org/10.1007/978-981-15-4814-7_15

execute immunization and hygiene and health practices in advance to promote prevention.

Keywords COVID 19 · Pandemic · Fears and Myths · Pandemic Preparedness · Pandemic stress · Children · School · Education Policy · Stress and depression in children

15.1 Introduction

Life seems to give lots of lessons to us every moment, but we fail to comprehend unless we face a catastrophe and till we try to understand the laws of nature, it is already too late. The current situation of the world is no less than a world war which has put everyone in a skeptical situation especially dealing with young and blank minds. The coronavirus disease 2019 (COVID-19) pandemic has created anxieties, turbulence, and fears in the young minds, and they barrage their parents with heaps of questions. COVID-19 has caused a global concern for Public Health Emergency. However, still the ongoing debate on the emergence of virus and its immediate host has proved its homology with bat coronavirus isolate RaTG13 (Li et al. 2020). It is responsible for the social and economic losses worldwide and is being transmitted at a very rapid pace. Health experts are still sailing in turbulent ocean to find an effective drug or vaccine. COVID-19-infected children may be asymptomatic or may present clinical symptoms such as fever, runny nose, dry cough, and fatigue including upper respiratory symptoms, congestion (Hong et al. 2020). Clinical reports also suggest poor immune response of the patients and are still trying to unveil the virus and its menaces (Weiss and Murdoch 2020). Along with the elderly population who are considered to be at the greater risk, even young children are not spared by the virus. According to the journal *Pediatrics*, nearly 2143 cases were reported in which 90% of children reported mild or moderate symptoms and were found to be at lower risk. But 6% of children showed intense and acute symptoms although none of the children died (Dong et al. 2020). Now the main concern was how to tackle the children and teach them to isolate themselves. Figure 15.1 explains the three categories of people and their role for not letting the infection escalate.

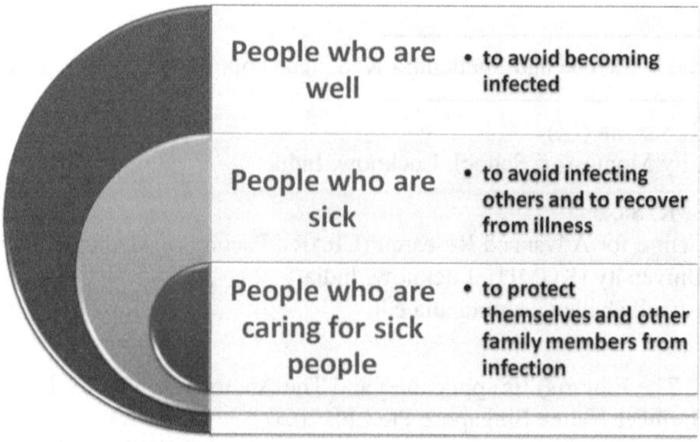

Fig. 15.1 Categories of people and how they can mitigate the infection of COVID-19

Children need to understand the above fact given in the figure more deeply because they usually love to be at crowded places such as schools, parks, etc. and love to play with their friends, so they have more chances to capture the infection and act as major disease transmitters and can intensify the pandemic. Due to lack of knowledge, the lockdown situations or imposing self-isolation on them by their parents may feel like a burden and is making them both apprehensive and restless. Either they are taking it just a vacation which can be enjoyed by numerous ways or they are in a dilemma about the repercussions of not following the rules. Children are always experimenting and believe only when the facts are presented in front of them in a way that they understand the meaning and necessity of refraining themselves from activities which can be the cause of concern.

The studies reveal that the children are equally susceptible to infection as the elderly people. Many other factors that increase their susceptibility include poverty, malnutrition, unawareness, and special healthcare needs, so the following topics will give us deeper insight into the children's perspective and the role of the people around them to combat the situation which may crop up due to this pandemic (Vaughan and Tinker 2009).

15.2 Fears and Myths in Children

Fears or anxieties are the usual part of the normal development of the child (Gullone 2000). But the myths and the pessimistic information surfacing around the children during the critical situations like COVID-19 pandemic may inoculate those fears in children which become a major health concern (Kessler et al. 2005). The way the parents transfer the information to the children about the pandemic is highly responsible for the fear levels of the children apart from the information received from media, pals, school and other community members, or direct encounter with the infection (Remmerswaal and Muris 2011). The haunting thoughts that may probably capture kids' minds right now are depicted in Fig. 15.2.

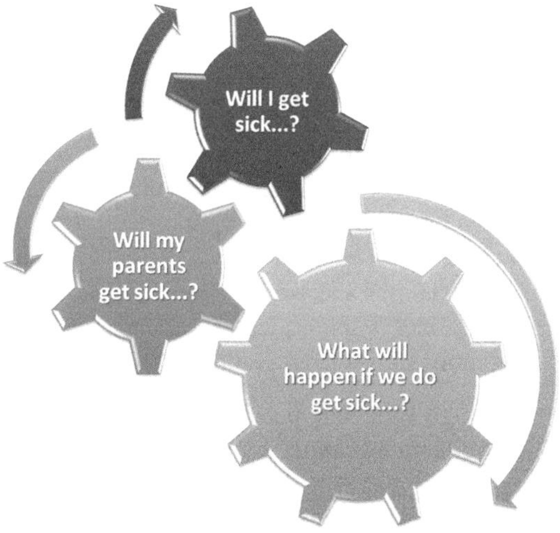

Fig. 15.2 The burning questions in the mind of children. These are the flickering questions in the minds of our kids due to the fear of pandemic

Away from their school, daycares, or routine outings, children are feeling helpless and discomfort for they do not have an idea that when would this mayhem be over and when they will return to normal days. Sans playgrounds and friends, they are in a state of frustration. For older kids it is quite easy because they have the maturity level to understand and level their anxieties. But the younger ones have lots of questions as to: Why so much of hand washing is required? what will happen if they get sick? when will they met their friends? and so on.

15.3 Children's Perspective

Children are keen observers; they can easily understand the agony, fear, and anxiety in their parents, peers, and community people (Liu 2020). Their perspective about the pandemic situation depends upon the behavior of the people around them. Hearing from social media, family discussions, and alarming news may make them more stressful, anxious, clingy, or uncertain in terms of their attitude. Schools being shut down, promotion of social distancing and parents working at home make them little apprehensive about the situation. Often children of lower age group are perplexed in such critical situations that why their parents have to work instead of playing with them or why they cannot play outside with their friends, etc.

15.4 Engaging Children for Pandemic Preparedness

When children are stressed, their bodies respond in a different way like by screaming, hiding, becoming sad, etc. To help them cope with these responses, it is important to acknowledge their feelings and involve them in different activities like:

- Reading books
- Engaging in arts and craft
- Helping parents in daily chores
- Exercise and Yoga to remain fit
- Meditation and spiritual talks to remain emotionally balanced
- Praying
- Practicing mindfulness
- Gardening to remain close to nature
- Searching online learning options
- Socializing on phone
- Writing stories and poems
- Playing indoor games with family members
- Listening stories from grandparents, etc.

Although pandemic phase is difficult to manage because we are throughout in a state of dilemma whether we will be safe survivors or not yet we have to tackle the

Fig. 15.3 A child's imagination and reflection about the novel coronavirus 2019. It shows that children are keen to know about the virus and are ready to face it. (©Picture courtesy: Aditi Saxena, Sixth Standard)

situation by our positive approach, so that our children should strengthen themselves. We can ask the children especially of lower age group to give their input by depicting their knowledge about the virus by writing stories and poems or by drawing posters, cartoons demonstrating their courage to fight the demon coronavirus as shown in Fig. 15.3. In addition, it is essential that families are not short of their basic needs (e.g., food, shelter, and clothing) otherwise instead of dealing with their children's stress, they will themselves be worrying about their inefficiency to provide proper nutrition and safety to their children.

15.5 Behavioral Management of Children (Prevention)

As per the information circulating through the media, the risk of exposure to COVID-19 is low in children as compared to adults. Immature children with less understanding of the situation are more susceptible to behavioral changes that disrupt their daily lives. Therefore, supporting them and protecting their behavior is necessary by people close to them not only during pandemics but also post pandemic. Children watch and interpret on their own and are easily carried away by the reactions of their parents, peers, and other community members, so we have to learn to guard our behavior first in order to prevent turbulence in budding minds.

15.6 Planning for School Closure

Community mitigation has been proved to be a good method to control disease spread. According to the report of CDC, some parts of the world have already experienced the 1918 influenza pandemic (Centers for Disease Control and Prevention n.d.). Previous experiences suggest that certain initiatives like closure of schools along with other precautions may help control the spread of pandemic in children.

Other mitigation planning such as avoiding large gatherings, minimizing or completely avoiding social gatherings or cancelling large events, and social distancing will prevent the spread of disease (Thorp 2020).

Disasters do not knock the doors before they come, they are unvoiced and walk into our lives without giving us time for preparation, so we should always be organized and ready for situations like school closures or complete lockdown. To limit the conduction of SARS-CoV-2 or any other health crisis, it is strictly required segregating the students and the teachers.

Advance and special planning for schools and training sessions can be conducted by health departments to implement effective measures, sanitation, and disease control guidelines (Stevenson et al. 2009). Imparting knowledge about health policies and programmers should be the part of school activities to promote prevention and control of bacterial or viral infections rather than only limited emphasis on immunization (Brener et al. 2007). These can also be reflected on school's portal throughout. However, many school days are lost due to shutting down of schools, but by effective planning, students can remain connected to their schools through Webinars, Google classrooms, etc. (Hardy 1991). Just smart planning and execution can overcome many problems.

15.7 Role of Teachers

It is the responsibility of the teachers to keep their students updated with the facts and attend to their queries in a calm and polite way. The fears and anxieties of the children can be reduced by providing scientific facts and sharing accurate information about COVID-19. It is not an easy task to make children understand the acuteness of the pandemic so easily. They can learn best when correct sequence of knowledge is imparted to them. Figure 15.4 shows the different levels of the methods which a teacher can adopt to make the children of lower age group realize easily who are difficult to be tackled under such pandemic situations.

Based on reputable sources such as UNICEF and the World Health Organization, the preschool children should be engaged in some activities to make them understand the protocols of their safety during any pandemic like hand washing, social distancing, avoiding handshakes, using handkerchief during cough and sneeze, etc. (Fisher and Wilder-Smith 2020). The secondary school children can be asked to form their own groups and portals to increase the awareness among their peers, families, and other community members through posters public demonstrations, discussions to encourage them to express and communicate their feelings.

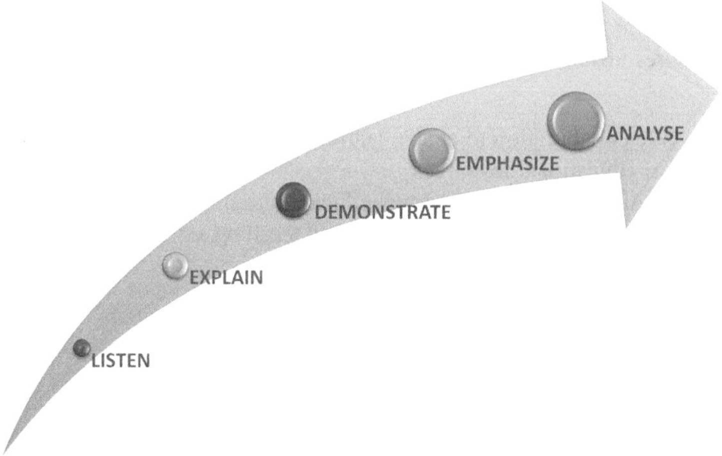

Fig. 15.4 Teaching levels for effective learning. A concerned teacher can use these techniques to unfold the fears and mysteries of immature learners

15.8 Role of Educational Institutes/Schools

Closing down of schools does not mean that their role is finished during any pandemic situation. Instead schools act as the major source to impart vital training to students and their families to help them understand the repercussions of violating the norms laid down by the government and health facilitators. School leaders, teachers, and support staff can focus on the well-being of students and prepare for the continuity of education, including working on online and remote learning options. Creating online classes and sharing and promoting e-learning can help students overcome the interruption in their academics (Rzymski and Nowicki 2020).

15.9 Role of Educational Policy Makers

About 80% of pupil population of the globe are affected by the closure of educational institutes due to the COVID-19 pandemic. Some of the students are deprived of the regular education for months which is being recognized as a major challenge by the governments and education policy makers. UNESCO is persistently observing and holding webinars, implicit meetings, and discussions with the educationists and the policy makers globally to meet out the challenges in education.

To diminish the interruption in studies during pandemics, many countries are implementing various forms of strategies, including Internet and distance learning, and are well prepared to impart education during any crisis (Wang et al. 2020).

Education policy makers can plan out in advance taking examples from early outbreaks like Ebola, SARS, and now COVID-19. For the continuation of education, some measures taken by the important educational organizations of the world may include:

(a) Worldwide collaboration to enable learning
(b) Sharing of educational resources
(c) Exchange of novel ideas and virtual education tools
(d) Teacher training to enhance e-learning and teaching skills
(e) Using distance learning
(f) Free access to educational sites
(g) Adapting existing educational sites for use in Smartphone
(h) Dealing with telecom companies to eliminate the cost of accessing material from a Ministry of Education site.

15.10 Role of a Parent

Most of the parents find no other option than to trust their doctors, nurses, government, and child care centers during any pandemic (King et al. 2018). The preparedness of a parent to deal with the children's anxiety depends on authenticity of the information they get.

During the COVID pandemic, distress felt by children of any age group is certainly due to school closure, distancing with friends, and sports, and the anxiety levels of the children can be diminished by extra love and care. While talking to the children, it is important that we should tell the truth to the children that if they follow the proper guidelines they will be alright instead of maximizing their fears just to make them sit quietly. It becomes necessary as a parent to rub off their fears and prevent creating a difficult scenario (Rich 2020). For this we need to understand the *state of mind* of the children as they may have:

1. Limited or no information or even misinformation
2. Fear and panic about the current pandemic
3. Stress and frustrated with their parents expressions and reactions
4. The advices for parents facing the wrath of their children include:

   (a) Control of their anxiety levels first.
   (b) Do not panic in the presence of children around the topic as our conduct cause ripples in their minds.
   (c) Introspection of self-mental status before facing the children.
   (d) Keep your fear, aggravation, and apprehension aside and prepare yourself for a fact-based conversation.
   (e) Try to find out what our children know about the virus and what are their misconceptions.
   (f) Lend your ears patiently before giving any suggestions or falling in any arguments.

(g) Then discuss about basic guidelines of prevention.
(h) Never load your child with your worries and future concerns.
(i) Tell them about the meaning and necessity of lockdown and self-isolation.
(j) Reassure your child that the concerned people are working round the clock and leaving no stone unturned to control the situation.
(k) Tell them that we can also support in bringing down the pandemic by playing our role wisely like distancing ourselves socially, maintaining routine hygiene and cleanliness, limiting travels, taking a balanced diet to avoid any other health issues, and take care of physical as well as mental well-being.
(l) Avoid watching or reading information causing panic and stress.
(m) Playing games, reading story books painting together, and doing household chores together can be followed.
(n) Making children sit and study is really a tough job. Let us make them learn and earn.
(o) We can give them a chance to earn money to buy their favorite toy or gadget in future by making them do what we want them to.
(p) We have to try our best to keep them busy, so that they remain calm and engaged during the pandemic to distress them and also provide them useful factual information to remain safe without anxiety.
(q) We have to explain our kids that COVID virus is very arrogant and egoistic; hence, it will not come to your house unless and until you go out to bring it home.

15.11 Dealing with Stress and Depression in Children During Pandemic Like Emergencies

COVID-19 has put the world's medical community to cooperate combat the virus. The countries as well as the global institutions need to put their hands together to help each other for the securing world peace and prosperity (UNESCO 2020).

To deal with the stress among children, various programmers need to be organized with the involvement of youth to understand their perspective and behavioral aspect to deal and confront future pandemics. Constructing a panel of different category of folks who can look into every aspect of the child imaginations and perspectives can effectively give wider solutions to for pandemic planning (Koller et al. 2010). This panel of intellectuals may include:

- Young volunteers
- Child psychologists
- Social reformers
- Counselors
- Educationists
- Medical representatives

We have to remember that together we can and will come out of any adverse situation.

15.12 Ensuring Mental Well-Being and Spiritual Upbringing

Mental well-being is also a part of other healthcare system especially during pandemics. Meditation, group prayers, reading spiritual and motivational stories, and singing spiritual or motivational songs can help out to release stress. The novel aspect of technology such as television, the Internet, and smartphone can also play an important role in delivering face to face support and training required during critical situations (Reisman et al. 2006). TED talks and spiritual lectures should be broadcast on regular basis to provide moral support.

State and local health departments should put their best foot forward to ensure that through media they arrange shows on mindfulness, talks with experts, and religious representative to uplift the mental status of children as well as adults. Social workers, child psychologists, psychiatrists, school psychologists, family counselors and therapists, and community organizations should collaborate to ensure mental health services.

15.13 Conclusions

Our understanding with the ongoing COVID-19 disaster shows that although children are at lower risk still mental trauma of the children should be dealt in a very effective way to strengthen them emotionally. Global organizations need to focus more on issues related to children by effective planning and should endorse necessary measures.

Right from the grassroot level involving home, community, local bodies, and national and international organizations all have to combat the pandemic jointly as we all have only one future our "children." We cannot afford to wait further for any adverse situation to formulate new guidelines. Whatever the earlier pandemics have taught us, we have to act according to that only. Disaster management bodies are doing their job round the clock. We have to ensure our children that they have to evolve as fighters in any disaster and come out as winners.

15.14 Future Perspectives

Hand on experience with disease outbreaks and pandemics opens the new conversation between policy makers, educationists, health administrators, law makers, social workers, and many more to look into the virtual pandemic which may be thrown by nature on humanity. How to deal with it in future is a major concern. Immunization policies at schools and colleges, health advisories to be laid down and followed, and immunization requirements need to be framed and prepared in advance. Few studies systematically investigate the range of factors that discuss the medical policies, facilitators, and obstacles in their implementation (Fawole et al. 2019).

References

Brener ND, Wheeler L, Wolfe LC, Vernon-Smiley M, Caldart-Olson L (2007) Health services: results from the School Health Programs and Policies Study 2006. J Sch Health 77:464–485. PubMed PMID: 17908103

Centers for Disease Control and Prevention (n.d.) Community strategy for pandemic influenza mitigation. http://www.pandemicflu.gov/plan/community/commitigation.html. Accessed 28 July 2009

Dong Y, Mo X, Hu Y, Qi X, Jiang F, Jiang Z, Tong S (2020) Epidemiological characteristics of 2143 pediatric patients with 2019 coronavirus disease in China. Pediatrics. https://doi.org/10.1542/peds.2020-0702. pii: e20200702. PubMed PMID: 32179660. [Epub ahead of print]

Fawole OA, Srivastava T, Feemster KA (2019) Student health administrator perspectives on college vaccine policy development and implementation. Vaccine 37(30):4118–4123. https://doi.org/10.1016/j.vaccine.2019.05.073. PubMed PMID: 31164307

Fisher D, Wilder-Smith A (2020) The global community needs to swiftly ramp up the response to contain COVID-19. Lancet. https://doi.org/10.1016/S0140-6736(20)30679-6. pii: S0140-6736(20)30679-6. PubMed PMID: 32199470. [Epub ahead of print]

Gullone E (2000) The development of normal fear: a century of research. Clin Psychol Rev 20(4):429–451. PubMed PMID: 10832548

Hardy AM (1991) Incidence and impact of selected infectious diseases in childhood. Vital Health Stat 10(180):1–22. PubMed PMID: 1746160

Hong H, Wang Y, Chung HT, Chen CJ (2020) Clinical characteristics of novel coronavirus disease 2019 (COVID-19) in newborns, infants and children. Pediatr Neonatol. https://doi.org/10.1016/j.pedneo.2020.03.001. pii: S1875-9572(20)30026-7. PubMed PMID: 32199864. [Epub ahead of print]

Kessler RC, Berglund P, Demler O, Jin R, Merikangas KR, Walters EE (2005) Lifetime prevalence and age-of-onset distributions of DSM-IV disorders in the National Comorbidity Survey Replication. Arch Gen Psychiatry 62(6):593–602

King CL, Chow MYK, Wiley KE, Leask J (2018) Much ado about flu: a mixed methods study of parental perceptions, trust and information seeking in a pandemic. Influenza Other Respir Viruses 12(4):514–521. https://doi.org/10.1111/irv.12547.Epub. PubMed PMID: 29437291; PubMed Central PMCID: PMC6005583

Koller D, Nicholas D, Gearing R, Kalfa O (2010) Paediatric pandemic planning: children's perspectives and recommendations. Health Soc Care Community 18(4):369–377. https://doi.org/10.1111/j.1365-2524.2009.00907.x. PubMed PMID: 20180866

Li C, Yang Y, Ren L (2020) Genetic evolution analysis of 2019 novel coronavirus and coronavirus from other species. Infect Genet Evol 82:104285. https://doi.org/10.1016/j.meegid.2020.104285. PubMed PMID: 32169673

Liu K (2020) How I faced my coronavirus anxiety. Science 367(6484):1398. https://doi.org/10.1126/science.367.6484.1398. PubMed PMID: 32193330

Reisman DB, Watson PJ, Klomp RW, Tanielien TL, Prior SD (2006) Pandemic influenza preparedness: adaptive responses to an evolving challenge. J Homeland Secur Emerg Manage 3:1–28

Remmerswaal D, Muris P (2011) Children's fear reactions to the 2009 Swine Flu pandemic: the role of threat information as provided by parents. J Anxiety Disord 25(3):444–449. https://doi.org/10.1016/j.janxdis.2010.11.008. PubMed PMID: 21159486

Rich M (2020) 6 ways parents can support their kids through the coronavirus disease (COVID-19) outbreak. https://www.unicef.org/coronavirus. Accessed 20 Mar 2020

Rzymski P, Nowicki M (2020) Preventing COVID-19 prejudice in academia. Science 367(6484):1313. https://doi.org/10.1126/science.abb4870. PubMed PMID: 32193314

Stevenson E, Barrios L, Cordell R, Delozier D, Gorman S, Koenig LJ, Odom E, Polder J, Randolph J, Shimabukuro T, Singleton C (2009) Pandemic influenza planning: addressing the needs of children. Am J Public Health 99(Suppl 2):S255–S260. https://doi.org/10.2105/AJPH.2009.159970. PubMed PMID: 19797738; PubMed Central PMCID: PMC4504394

Thorp HH (2020) Time to pull together. Science 367(6484):1282. https://doi.org/10.1126/science.abb7518. PubMed PMID: 32179702

UNESCO (2020) Amidst global school closures UNESCO's Futures of Education initiative receives strong support. https://en.unesco.org/news/amidst-global-school-closures-unescos-futures-education-initiative-receives-strong-support. Accessed 20 Mar 2020

Vaughan E, Tinker T (2009) Effective health risk communication about pandemic influenza for vulnerable populations. Am J Public Health 99(Suppl 2):S324–S332. https://doi.org/10.2105/AJPH.2009.162537. PubMed PMID: 19797744; PubMed Central PMCID: PMC4504362

Wang G, Zhang Y, Zhao J, Zhang J, Jiang F (2020) Mitigate the effects of home confinement on children during the COVID-19 outbreak. Lancet 395(10228):945–947. https://doi.org/10.1016/S0140-6736(20)30547-X

Weiss P, Murdoch DR (2020) Clinical course and mortality risk of severe COVID-19. Lancet. https://doi.org/10.1016/S0140-6736(20)30633-4. pii: S0140-6736(20)30633-4. PubMed PMID: 32197108. [Epub ahead of print]

Chapter 16
Coping with Mental Health Challenges During COVID-19

Sujita Kumar Kar, S. M. Yasir Arafat, Russell Kabir, Pawan Sharma, and Shailendra K. Saxena

Abstract The ongoing pandemic of COVID-19 is a global challenge which resulted in significant morbidity and mortality worldwide. It has also adversely affected the economy and social integrity. There is rising concern about the mental health challenges of the general population, COVID-19-infected patients, close contacts, elderly, children and health professionals. This chapter focusses on various mental health challenges during the COVID-19 pandemic.

Keywords COVID-19 · Mental health · Coping · Pandemic

Sujita Kumar Kar and Shailendra K. Saxena contributed equally as first author.

S. K. Kar (✉)
Department of Psychiatry, Faculty of Medicine, King George's Medical University (KGMU), Lucknow, India

S. M. Yasir Arafat
Department of Psychiatry, Enam Medical College and Hospital, Dhaka, Bangladesh

R. Kabir
School of Allied Health, Faculty of Health, Education, Medicine, and Social Care, Anglia Ruskin University, Chelmsford, UK

P. Sharma
Department of Psychiatry, Patan Academy of Health Sciences, Arogin Health Care and Research Center, Kathmandu, Nepal

S. K. Saxena
Centre for Advanced Research (CFAR), Faculty of Medicine, King George's Medical University (KGMU), Lucknow, India
e-mail: shailen@kgmcindia.edu

S. K. Saxena (ed.), *Coronavirus Disease 2019 (COVID-19)*, Medical Virology: from Pathogenesis to Disease Control, https://doi.org/10.1007/978-981-15-4814-7_16

16.1 Introduction

There is a major health crisis going on in the world currently. A newly emerged zoonotic viral infection known as novel coronavirus disease (COVID-19) is affecting people, globally taking the form of a pandemic. Over the past few months, there is a significant increase in mortality and morbidity due to this pandemic. Till date (29 March 2020), there are more than 680,000 total cases with 31,920 deaths, 146,396 recovered over 202 countries (COVID-19 Coronavirus Pandemic 2020). As per the situation report of the World Health Organization (WHO), by the end of 28 March 2020, more than half of the global deaths and infected cases were from the European region (World Health Organization 2020a). As the disease is spreading in a rapid pace, most of the affected countries are not able to meet the demands of the personal protective equipment (PPE) and infrastructure requirement (World Health Organization 2020a). At the current stage, the major objectives laid by the WHO are prevention of human-to-human transmission, limiting the spread of infection to close contacts and medical professionals, preventing the development of complications in infected persons, isolation and quarantine facility provision, availing diagnostic and laboratory facility, research to produce specific treatment and vaccine and minimizing the socioeconomic impact on the community (World Health Organization 2020a). It has been noticed over the past few months that during this outbreak of COVID-19 infection, there are increasing mental health issues among the general population, elderly, children, migrant workers and healthcare professionals other than the patients with COVID-19 infection (Duan and Zhu 2020; Chen et al. 2020; Liem et al. 2020; Yang et al. 2020a, b). To date there are no specific recommendations from international bodies regarding addressing the mental health issues during this COVID-19 pandemic.

16.2 Impact of COVID-19 in Society

The global impact of COVID-19 has been profound, and the public health threat due to this is the most serious seen since the 1918 H1N1 influenza pandemic. The overall case fatality rate of COVISD-19 was 2.3% in China and could be variable in different countries (Novel Coronavirus Pneumonia Emergency Response Epidemiology Team 2020; Livingston and Bucher 2020). A nationwide analysis done in China showed that comorbidities are present in around one-fourth of patients with COVID-19 and predispose to poorer clinical outcomes (Novel Coronavirus Pneumonia Emergency Response Epidemiology Team 2020). The impact of the disease is beyond mortality, and morbidity has become apparent since the outbreak of the pandemic. A large population throughout the world is certain to have a massive psychological impact as evidenced by a preliminary report from China where among 1210 respondents more than half of the respondents rated the psychological impact as moderate-to-severe and about one-third reported moderate-to-severe anxiety

(Wang et al. 2020a). Studies post-SARS pandemic or post-Ebola indicate that even after recovering physically from the disease, individuals suffered from social and psychological problems and similar could be the impact with this pandemic (Bobdey and Ray 2020). Evidence suggests that vulnerable groups who are confined to their homes during a pandemic can have negative health outcomes. Children especially become physically less active and have much longer screen time, irregular sleep patterns and less favourable diets, resulting in weight gain and a loss of cardiorespiratory fitness (Wang et al. 2020b). Also, there are other direct and indirect implications of the closure of schools like unintended childcare obligations, which are particularly large in healthcare occupations (Bayham and Fenichel 2020). This could be related to the current situation in most of the countries throughout the world not only in child care but also in the adult and geriatric population (Heckman et al. 2020).

COVID-19 is a supply shock and a demand shock. Both the aspects will impact on aggregate trade flow (Baldwin and Tomiura 2020). It has both direct and indirect economic implications. The stocks and flow of physical and financial assets are interrupted. An increase in health budget and a lowering of overall GDP is sure to impact the whole world (McKibbin and Fernando 2020). Another area of impact would be travel and tourism. In the current scenario, the travel of any citizen of any country has been virtually been stopped. Also, even once the pandemic is over, it is almost certain to take a long time before people become confident of travel (Anzai et al. 2020; Dinarto et al. 2020).

Stigma and fear are other aspects of the outbreak of a pandemic. It can present major barriers against healthcare seeking, social marginalization, distrust in health authorities and distortion of public perceptions of risk, resulting in mass panic among citizens and the disproportionate allocation of healthcare resources by politicians and health professionals (Barrett and Brown 2008).

Impact on the sports and other mass gatherings throughout the world cannot be ignored (Gallego et al. 2020). Within weeks of the emergence of this pandemic in China, there have been circulation of misinformation, misleading rumours and conspiracy theories about the origin paired with fear mongering, racism and compulsive buying and stocking of goods and face masks. This can be attributed to impact the social media has created (Depoux et al. 2020). Over all the pandemic will have impact in all domains of the current world starting from health, society and economy and would also impact the future policy making at global, regional and country level (Djalante et al. 2020).

16.3 Emerging Mental Health Issues in COVID-19 Pandemic

The COVID-19 pandemic is a global emergency situation while the diagnosis of specific disorders needs a specific time period which is a major constraint to quantify the mental health issues. Moreover, many of the survivors may develop mental disorders long after the event. Therefore, multiple and complex confounding variables makes the issue hazy. Fortunately, studies evaluating the mental health issue have been coming out gradually which needs more time certainly to get replicable findings.

16.3.1 Among General Population

As the COVID-19 pandemic has been spreading rapidly across the globe, the foremost mental health issue has raised the level of stress or anxiety expressed in public mental health term (Dong and Bouey 2020). Inadequate knowledge regarding the incubation period of the virus, route of transmission, treatment and safety measures cause fear and anxiety (Li et al. 2020; Ho et al. 2020; Goyal et al. 2020). The locked-down state bounds residents to become homebound which causes negative mental health outcomes like anxiety states and insecurity regarding the future (Li et al. 2020). The citizens also feel monotony, disappointment and irritability under the locked-down state (Ho et al. 2020). One study reported severe and wide spectrum mental health impacts of the pandemic (Goyal et al. 2020). The event can precipitate new mental disorders and exacerbate the previously present disorders (Goyal et al. 2020). The general population can experience fear and anxiety of being sick or dying, helplessness, blame the people who are already affected and precipitate the mental breakdown (Goyal et al. 2020). A wide range of psychiatric disorders can be found such as depressive disorders, anxiety disorders, panic disorder, somatic symptoms, self-blame, guilt, posttraumatic stress disorder (PTSD), delirium, psychosis and even suicide (Goyal et al. 2020; Yi et al. 2020).

16.3.2 Among COVID-19 Cases

The suspected and/or confirmed COVID-19 persons largely experience fear regarding the high contagiousness and fatality (Wang et al. 2020a; Li et al. 2020). The quarantined people feel boredom, loneliness, anger, depression, anxiety, denial, despair, insomnia, harmful substance use, self-harm and suicidality (Wang et al. 2020a; Dong and Bouey 2020; Li et al. 2020; Yi et al. 2020). The survivors are the high-risk people to develop a wide range of mental disorders such as depression, anxiety and PTSD (World Health Organization 2020a). As a continuation of safety behaviours, patients may develop obsessive-compulsive disorder (OCD) (Li et al. 2020). Moreover, physical symptoms of COVID-19 such as fever, hypoxia and cough along with adverse effects of prescribed medications (corticosteroids) may cause more anxiety and mental distress (Wang et al. 2020a). A recent study of 1210 participants from 194 cities in China reported that 53.8% had a moderate or severe psychological impact, 31.3% had some sort of depression, 36.4% had some sort of anxiety and 32.4% had some sort of stress (Liu et al. 2020). Poor or very poor self-rated health status was significantly associated with a greater psychological impact of the COVID-19 (Liu et al. 2020).

16.3.3 Among Family Members and Close Contacts

Along with the persons with COVID-19, the family members and close contacts face psychological problems as they have been traced, isolated or quarantined which makes people anxious and guilty regarding the aftermath of the contagion, quarantine and stigma on their family members and friends (Wang et al. 2020a). The family members who lose their loved ones from the pandemic results in anger and resentment (Goyal et al. 2020). Furthermore, they also feel shame, guilt or stigma for those family members who are sick and/or quarantined, and some studies reported PTSD and depression among the family members and close contacts (Goyal et al. 2020). On the other hand, the children who have been isolated or quarantined during the pandemic have higher chances to develop acute stress disorder, adjustment disorder and grief (Shah et al. 2020). PTSD was reported among 30% of the children and early loss of or separation from parents during childhood also has long-term adverse effects on mental health, including higher chances of developing mood disorders, psychosis and suicidality (Shah et al. 2020).

16.3.4 Among Healthcare Workers

As pandemics are the global public mental health emergency, healthcare services demand increases sharply. Furthermore, many countries do not have adequate manpower as well as resources to cope with COVID-19. Thus, healthcare providers have to face an increased workload with the fear of being infected. Many times, they have been quarantined frequently when they contact COVID-19-confirmed persons.

Increased workload, isolation and discrimination are common which result in physical exhaustion, fear, emotional disturbance and sleep disorders (Ho et al. 2020). A recent study involving 1563 health professionals reported that more than half (50.7%) of the participants reported depressive symptoms, 44.7% anxiety and 36.1% sleep disturbance (Ho et al. 2020). Moreover, there are not adequate services to provide counselling and psychiatric screening services for anxiety, depression and suicidality for physicians who have been dealing with infected persons (World Health Organization 2020b). It is also meaningful to postulate that many physicians develop PTSD, depression, anxiety and burnout after the cessation of the pandemic (World Health Organization 2020b). Along with the physicians, the frontline healthcare providers (FHCP) can develop mental disorders such as depression, anxiety and PTSD (Li et al. 2020). Previous articles reported that FHCP (paramedics, ambulance personnel and healthcare workers) have also shown heightened stress and emotional disturbances and have higher levels of depression and anxiety (Goyal et al. 2020).

This is estimated as the chances of getting infected is much higher with the risk of exposure which creates a fear of transmission to their loved ones and children. Furthermore, the conflict professionalism and personal fear for oneself causes burnouts and physical and mental symptoms (Goyal et al. 2020).

16.3.5 Among Special Population (Old Age and Co-morbidities)

As this pandemic has been spreading rapidly across the world, it is bringing a considerable degree of fear, worry and concern among few certain groups particularly, in older adults and people with underlying comorbid disorders (Dong and Bouey 2020). It has a potential impact on the existing diseases, and the affected persons may lead to psychiatric symptoms which possibly related to the interplay of mental disorders and immunity (World Health Organization 2020b). The symptoms of COVID-19 can also worsen cognitive distress and anxiety among people who have poor mental capabilities previously (World Health Organization 2020b).

Patients with pre-existing severe mental illness (SMI) have been inevitably affected by the pandemic (Ho et al. 2020). In-patients, especially those requiring long-term hospitalization in closed wards, pose a high risk of cluster contagion. Due to traffic restrictions and isolation measures, outpatients with SMI are facing difficulties to receive maintenance treatment and may thus end up with mental relapse and uncontrollable situations (Ho et al. 2020). Patients with chronic physical illness (e.g., chronic renal failure, diabetes mellitus and cardio-cerebrovascular diseases) also need regular follow-up in hospitals which become problematic and raise the chances of deterioration.

16.4 Coping with Mental Health Issues During COVID-19 Pandemic

While the healthcare sector and government officials from all over the world is focusing on the control of the pandemic adopting various preventive strategies, there is little attention provided to the mental health status of the isolated, panicked and house-arrested people. Due to lack of regular social activities and staying at home for a longer time will impact their emotional well-being. Research has also shown that sudden outbreak can worsen the mental health conditions of those with pre-existing mental health illness (Ho et al. 2020).

To avoid a distressing situation, individuals should not get exposed to media coverage too much, to maintain a healthy relationship, get in touch with friends and

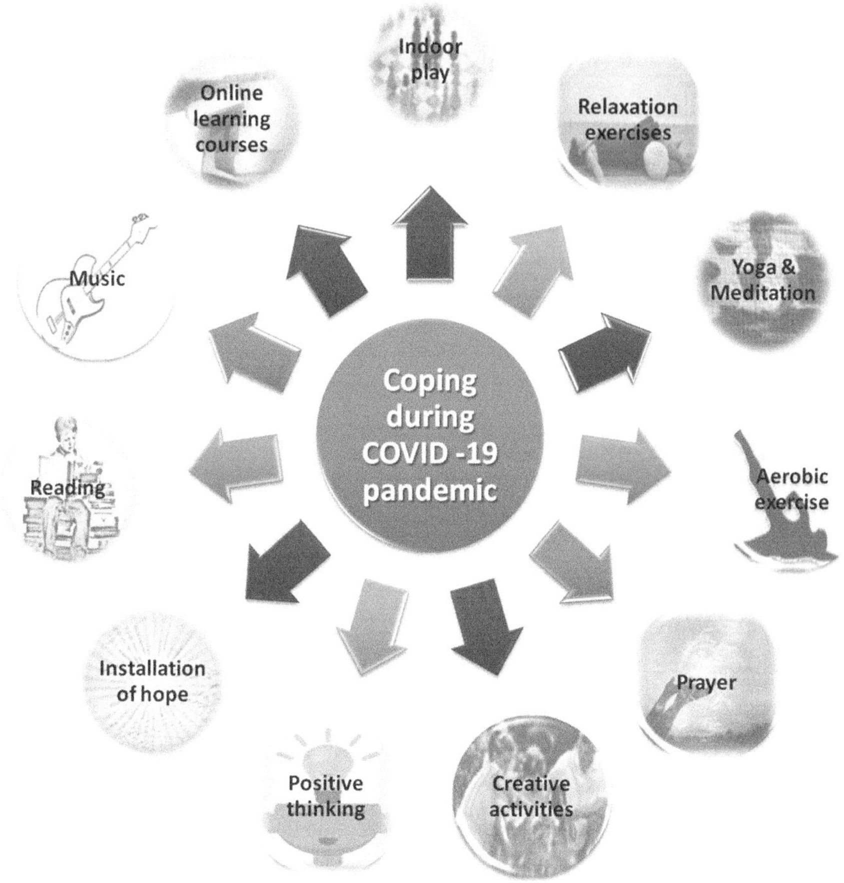

Fig. 16.1 Summary of coping measures during COVID-19 pandemic

family members on a regular interval using social media and start thinking positively (CDC 2020). If coronavirus anxiety shows up, try to share the fear with others, which will calm the fear, and also try to increase self-awareness by getting adequate sleep, exercising regularly and employing different relaxation techniques (Kecmanovic 2020). As recommended by Ho et al. (2020) in this era of technology, healthcare services can introduce providing online psychological support services for those individuals who lost their close relatives due to COVID-19 (Ho et al. 2020). To support the morale and mental health of the frontline healthcare professionals, healthcare organizations should introduce shorter working periods, regular breaks and rotating shifts (Ho et al. 2020). People can cope with the mental health challenges by adopting various lifestyle-related measures (Figs. 16.1 and 16.2).

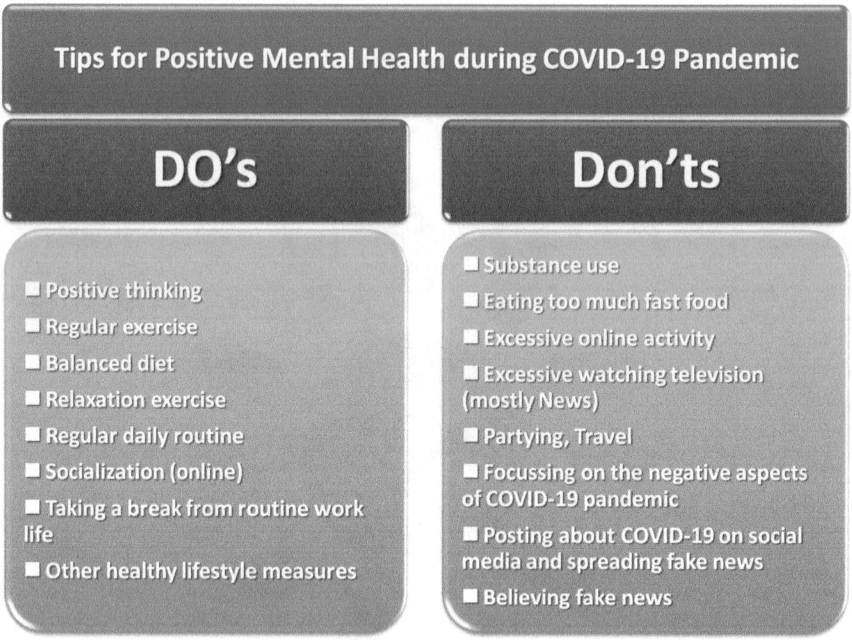

Fig. 16.2 Tips for positive mental health during COVID-19 pandemic

16.5 Myths and Facts About COVID-19

During a new infectious disease outbreak, a great deal of uncertainty remains from the pattern of transferring, risk factors involved and prevention and treatment (Schuchat et al. 2011). Rumours and myths create more panic among the general public as they are malevolent in nature and can alter people's observations towards the disease. The world is witnessing the same for the new public health crises with the emergence and spread of 2019 novel coronavirus. When the disease originated from Wuhan city of China, it was declared as a second-class infectious disease, but most of the areas of the country adopted the first level of response measure to control, and the measures were taken has no scientific basis and no effective outcomes were recorded after applying those measures by the China government (Xiao and Torok 2020).

The virus cannot be killed by cold and snow, and it can be transmitted in areas with hot and humid climate (WHO 2020). People of all age groups are susceptible to get infected with COVID-19. Elderly people with underlying health conditions such as diabetes, heart disease and asthma are more vulnerable (Fong 2020). Although there is no significant number of paediatric cases so far, children are vulnerable to the infection (Hong et al. 2020), and to date, there is no evidence of vertical transmission of this infection (Baud et al. 2020).

Several national and international newspapers, tabloids and media channels all over the world are reporting that the smokers are prone to catch coronavirus infection due to weakened lungs and will put the smokers at risk (Mullin 2020). However, a recent systematic review finding revealed that no significant association is found between active smoking and the severity of COVID 19 (Lippi and Henry 2020).

There are reports of using oseltamivir, lopinavir/ritonavir, prednisone, antibiotics, and traditional Chinese medicine for the treatment of patients with COVID-19. Again, there is no scientific evidence to support that they will be effective against COVID-19 apart from scrupulous personal care such as the use of personal protection precaution to reduce the risk of transmission, early diagnosis, isolation and supportive treatments for affected patients (Xiao and Torok 2020; Wang et al. 2020c).

There is also some misconception among the general people that by taking hot bath, people will not get infected with the infection or spraying alcohol or chlorine all over the body can kill the infection (WHO 2020). Proper public health information should be provided, based on scientific research to general people to reduce stress and anxiety, otherwise it will be difficult to implement control measures.

16.6 Precautionary Measures and Recommendations

No definite treatment is available for the treatment of the COVID-19 infection. Prevention is the best strategy to combat the COVID-19 pandemic. Prevention is not a difficult task as it is commonly thought to be. For the effective prevention of COVID-19, broadly two types of precautionary measures to be taken, as mentioned below:

1. General precautionary measures (Fig. 16.3): It is meant for everybody in the community.
2. Specific precautionary measures (Fig. 16.4): It is meant for persons who are sick, close contacts of COVID-19, travellers and healthcare workers.

Broadly, there are three groups of population as mentioned below:

(a) General population
(b) COVID-19 cases and close contacts
(c) Healthcare workers

The precautions and recommendations are targeted to address the needs of the above three groups of the population. Prevailing myths and unawareness about precautionary measures may cause distress among people. There is a need to follow certain recommendations for effective coping with mental health challenges (Table 16.1).

Individuals who experience psychological distress must report or inform their difficulties, rather than hiding them. Individuals who experience persistence distress

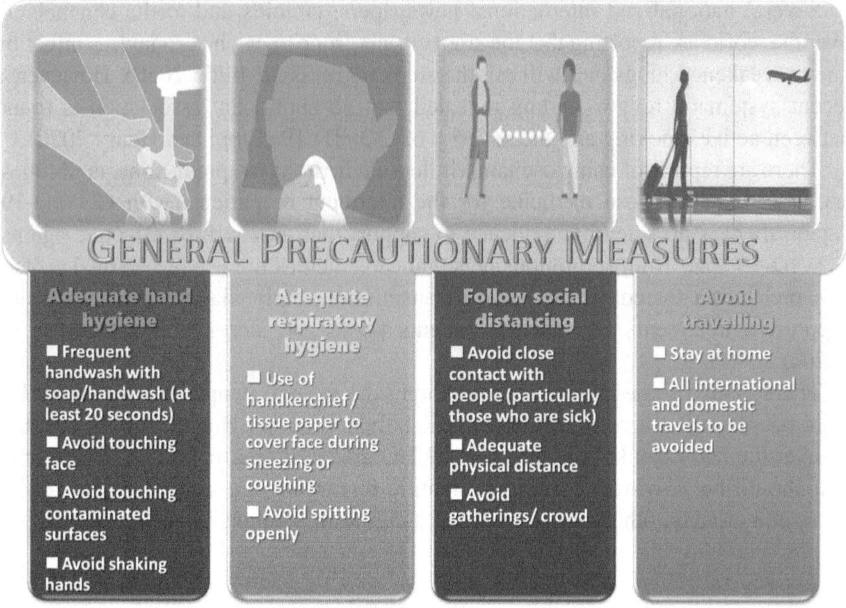

Fig. 16.3 General precautionary measures during COVID-19 pandemic. (Source: Adapted from Centre for Disease Control and Prevention, USA)

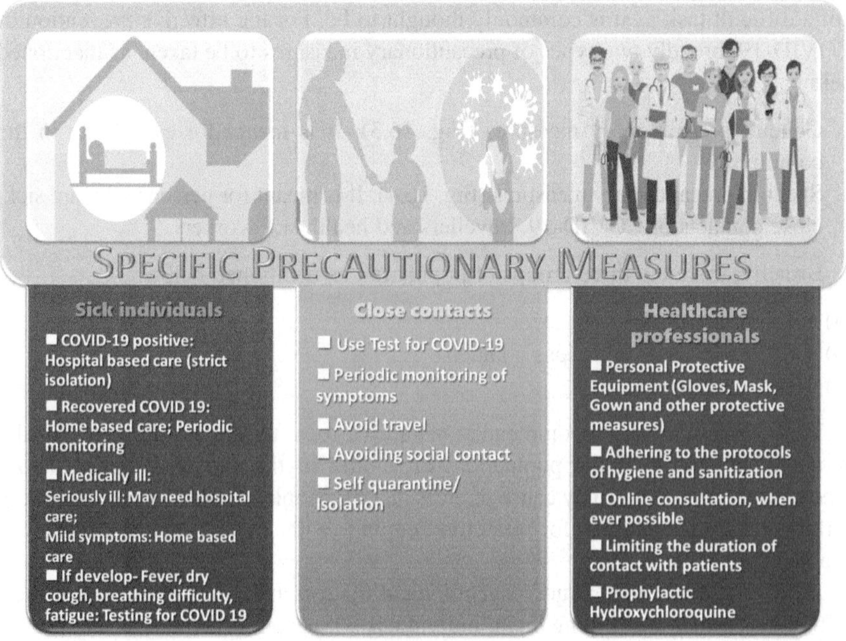

Fig. 16.4 Specific precautionary measures during COVID-19 pandemic. (Source: Adapted from Centre for Disease Control and Prevention USA)

Table 16.1 Recommendations for effective coping with mental health challenges

1. Adequate awareness about the COVID-19 and regular updates (as understanding about COVID-19 changing day by day) about appropriate precautionary measures
2. Developing preparedness to meet the challenges like scarcity of resources
3. Ignoring fake news and social media posts that spreads panic
4. Regular scheduling of the daily activities
5. Inclusion of indoor recreational activities and relaxation exercises to daily practice
6. Approaching (rather than avoiding) healthcare system, if any symptoms develop
7. Positive thinking and installation of hope

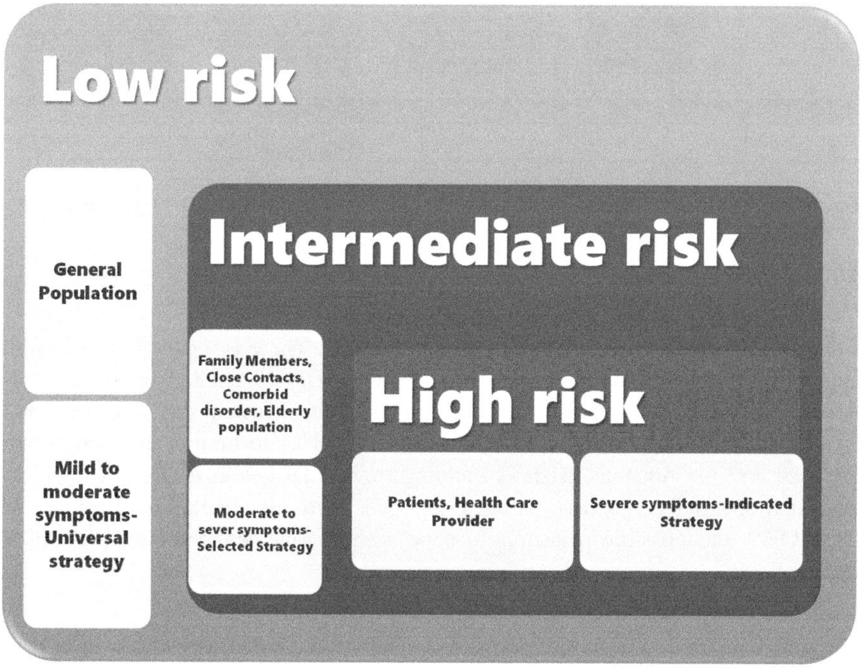

Fig. 16.5 Recommendations according the COVID-19 risk severity

may seek help from the mental health professionals through helplines available or in hospitals in cases of emergency situations. Figures 16.5 and 16.6 summarize the recommendations according the risk severity and management approach to mental health difficulties during COVID-19 pandemic, respectively.

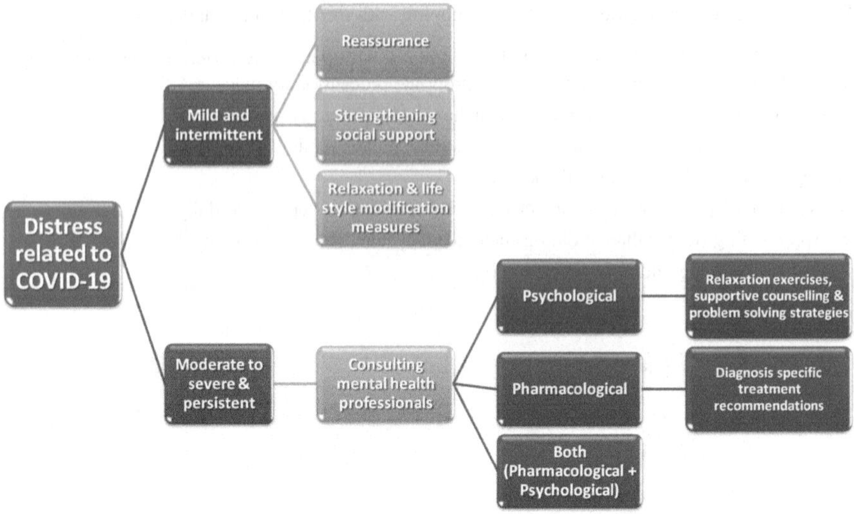

Fig. 16.6 Management approach to mental health difficulties during COVID-19 pandemic

16.7 Conclusions

COVID-19 carries significant mental health hazards. There is a paucity of research addressing the mental health issues during the COVID-19 pandemic. As the mortality and morbidity statistics are reaching new peaks every day, isolation and lockdown states are getting prolonged, recreational opportunities for people are lessened and the financial crisis is building in, mental health issues are likely to grow exponentially. There is a need to understand the mental perspectives of COVID-19 and possible measures to cope with the pandemic for their effective management.

16.8 Future Perspectives

The mental health issues associated with the COVID-19 pandemic can be immediate (short-term) or remote (long-term). Existing literature addresses the immediate mental health concerns only. It is important to see the long-term mental health sequels of COVID-19 infection. Earlier pieces of evidence suggest that maternal exposure to influenza infection during the epidemic of influenza in Europe increased the risk of schizophrenia in offspring, possibly by altering the neurodevelopmental process (Mednick et al. 1988; Murray et al. 1992). Similarly, childhood exposure to measles may later result in the development of subacute sclerosing panencephalitis (SSPE) (Campbell et al. 2007). Nothing is known about the after-effects of novel coronavirus infection; hence, there is a need for extensive research in terms of its impact on various groups of populations (pregnant, young children, adults and other vulnerable populations).

Similarly, it is required to understand the mental healthcare needs of patients with COVID-19, close contacts, health professionals dealing with COVID-19 patients and the general population. Future research should also consider the feasibility and efficacy of various online psychotherapeutic interventions during the COVID-19 pandemic, globally with a specific focus in the low- and middle-income countries (LMICs). As there is a concern about contacting infection during direct contact with patients, online consultation can be a potential mode of delivering therapy (Greenhalgh et al. 2020).

Executive Summary

- Mental health issues differ among various populations during the COVID-19 pandemic.
- Vulnerable populations like COVID-19 cases, close contacts, elderly, children and health professionals are expected to have more difficulties with coping.
- Appropriate precautionary measures may reduce the psychological distress.
- Myths associated with COVID-19 may also lead to distress and inappropriate lifestyle measures.
- People experiencing distress should adopt various healthy relaxation measures and if required help from mental health professionals.

References

Anzai A, Kobayashi T, Linton NM, Kinoshita R, Hayashi K, Suzuki A et al (2020) Assessing the impact of reduced travel on exportation dynamics of novel coronavirus infection (COVID-19). J Clin Med 9(2):601

Baldwin R, Tomiura E (2020) 5 Thinking ahead about the trade impact of COVID-19. Econ Time COVID-19. 59

Barrett R, Brown PJ (2008) Stigma in the time of influenza: social and institutional responses to pandemic emergencies. J Infect Dis 197(Suppl 1):S34–S37

Baud D, Giannoni E, Pomar L, Qi X, Nielsen-Saines K, Musso D et al (2020) COVID-19 in pregnant women—authors' reply. Lancet Infect Dis. https://doi.org/10.1016/S1473-3099(20)30192-4

Bayham J, Fenichel EP (2020) The impact of school closure for COVID-19 on the US Healthcare Workforce and the Net Mortality Effects. medRxiv. 2020.03.09.20033415

Bobdey S, Ray S (2020) Going viral–Covid-19 impact assessment: a perspective beyond clinical practice. J Mar Med Soc 22(1):9

Campbell H, Andrews N, Brown KE, Miller E (2007) Review of the effect of measles vaccination on the epidemiology of SSPE. Int J Epidemiol 36(6):1334–1348

CDC (2020) Mental health and coping during COVID-19. http://adultmentalhealth.org/mental-health-and-coping-during-covid-19/. Cited 29 Mar 2020

Chen Q, Liang M, Li Y, Guo J, Fei D, Wang L et al (2020) Mental health care for medical staff in China during the COVID-19 outbreak. Lancet Psychiatry 7(4):e15–e16

COVID-19 Coronavirus Pandemic (2020). https://www.worldometers.info/coronavirus/. Cited 29 Mar 2020

Depoux A, Martin S, Karafillakis E, Bsd RP, Wilder-Smith A, Larson H (2020) The pandemic of social media panic travels faster than the COVID-19 outbreak. J Travel Med. https://doi.org/10.1093/jtm/taaa031

Dinarto D, Wanto A, Sebastian LC (2020) Global health security – COVID-19: impact on Bintan's tourism sector. https://dr.ntu.edu.sg//handle/10356/137356. Cited 29 Mar 2020

Djalante R, Shaw R, DeWit A (2020) Building resilience against biological hazards and pandemics: COVID-19 and its implications for the Sendai Framework. Prog Disaster Sci 6:100080

Dong L, Bouey J (2020) Public mental health crisis during COVID-19 pandemic, China. Emerg Infect Dis 26(7). https://doi.org/10.3201/eid2607.200407

Duan L, Zhu G (2020) Psychological interventions for people affected by the COVID-19 epidemic. Lancet Psychiatry 7(4):300–302

Fong LY (2020) Frequently asked questions and myth busters on COVID-19. https://worldwide.saraya.com/about/news/item/frequently-asked-questions-and-myth-busters-on-covid-19. Cited 29 Mar 2020

Gallego V, Nishiura H, Sah R, Rodriguez-Morales AJ (2020) The COVID-19 outbreak and implications for the Tokyo 2020 Summer Olympic Games. Travel Med Infect Dis:101604

Goyal K, Chauhan P, Chhikara K, Gupta P, Singh MP (2020) Fear of COVID 2019: first suicidal case in India. Asian J Psychiatry 49:e101989

Greenhalgh T, Wherton J, Shaw S, Morrison C (2020) Video consultations for covid-19. BMJ 368: m998

Heckman GA, Saari M, McArthur C, Wellens NI, Hirdes JP (2020) RE: COVID-19 response and chronic disease management. https://www.cmaj.ca/content/re-covid-19-response-and-chronic-disease-management. Cited 29 Mar 2020

Ho CS, Chee CY, Ho RC (2020) Mental health strategies to combat the psychological impact of COVID-19 beyond paranoia and panic. Ann Acad Med Singap 49(1):1

Hong H, Wang Y, Chung H-T, Chen C-J (2020) Clinical characteristics of novel coronavirus disease 2019 (COVID-19) in newborns, infants and children. Pediatr Neonatol. https://doi.org/10.1016/j.pedneo.2020.03.001

Kecmanovic J (2020) 7 science-based strategies to cope with coronavirus anxiety. The Conversation. http://theconversation.com/7-science-based-strategies-to-cope-with-coronavirus-anxiety-133207. Cited 30 Mar 2020

Li W, Yang Y, Liu Z-H, Zhao Y-J, Zhang Q, Zhang L et al (2020) Progression of mental health services during the COVID-19 outbreak in China. Int J Biol Sci 16(10):1732–1738

Liem A, Wang C, Wariyanti Y, Latkin CA, Hall BJ (2020) The neglected health of international migrant workers in the COVID-19 epidemic. Lancet Psychiatry 7(4):e20

Lippi G, Henry BM (2020) Active smoking is not associated with severity of coronavirus disease 2019 (COVID-19). Eur J Intern Med. https://doi.org/10.1016/j.ejim.2020.03.014

Liu JJ, Bao Y, Huang X, Shi J, Lu L (2020) Mental health considerations for children quarantined because of COVID-19. Lancet Child Adolesc Health. https://www.thelancet.com/journals/lanchi/article/PIIS2352-4642(20)30096-1/abstract. Cited 30 Mar 2020

Livingston E, Bucher K (2020) Coronavirus disease 2019 (COVID-19) in Italy. JAMA. https://jamanetwork.com/journals/jama/fullarticle/2763401. Cited 29 Mar 2020

McKibbin WJ, Fernando R (2020) The global macroeconomic impacts of COVID-19: seven scenarios. Social Science Research Network, Rochester, NY. Report No.: ID 3547729. https://papers.ssrn.com/abstract=3547729. Cited 29 Mar 2020

Mednick SA, Machon RA, Huttunen MO, Bonett D (1988) Adult schizophrenia following prenatal exposure to an influenza epidemic. Arch Gen Psychiatry 45(2):189–192

Mullin G (2020) SMOKE SCREEN Smoking 'may increase risk of catching coronavirus' and worsen symptoms, scientists warn. https://www.thesun.co.uk/news/11098194/smoking-increase-coronavirus-risk-worsen-symptoms/. Cited 29 Mar 2020

Murray RM, Jones P, O'Callaghan E, Takei N, Sham P (1992) Genes, viruses and neurodevelopmental schizophrenia. J Psychiatr Res 26(4):225–235
Novel Coronavirus Pneumonia Emergency Response Epidemiology Team (2020) [The epidemiological characteristics of an outbreak of 2019 novel coronavirus diseases (COVID-19) in China]. Zhonghua Liu Xing Bing Xue Za Zhi Zhonghua Liuxingbingxue Zazhi 41(2):145–151
Schuchat A, Bell BP, Redd SC (2011) The science behind preparing and responding to pandemic influenza: the lessons and limits of science. Clin Infect Dis 52(Suppl 1):S8–S12. https://academic.oup.com/cid/article/52/suppl_1/S8/498182. Cited 30 Mar 2020
Shah K, Kamrai D, Mekala H, Mann B, Desai K, Patel RS (2020) Focus on mental health during the coronavirus (COVID-19) pandemic: applying learnings from the past outbreaks. Cureus 12(3). https://www.cureus.com/articles/29485-focus-on-mental-health-during-the-coronavirus-covid-19-pandemic-applying-learnings-from-the-past-outbreaks. Cited 30 Mar 2020
Wang C, Pan R, Wan X, Tan Y, Xu L, Ho CS et al (2020a) Immediate psychological responses and associated factors during the initial stage of the 2019 coronavirus disease (COVID-19) epidemic among the general population in China. Int J Environ Res Public Health 17(5):1729
Wang G, Zhang Y, Zhao J, Zhang J, Jiang F (2020b) Mitigate the effects of home confinement on children during the COVID-19 outbreak. Lancet 395(10228):945–947
Wang D, Hu B, Hu C, Zhu F, Liu X, Zhang J et al (2020c) Clinical characteristics of 138 hospitalized patients with 2019 novel coronavirus–infected pneumonia in Wuhan, China. JAMA. https://doi.org/10.1001/jama.2020.1585
WHO (2020) Coronavirus disease (COVID-19) advice for the public: myth busters. World Health Organization. https://www.who.int/emergencies/diseases/novel-coronavirus-2019/advice-for-public/myth-busters. Cited 29 Mar 2020
World Health Organization (2020a) Coronavirus disease 2019 (COVID-19) situation report—68. World Health Organization. https://www.who.int/docs/default-source/coronaviruse/situation-reports/20200328-sitrep-68-covid-19.pdf?sfvrsn=384bc74c_2. Cited 29 Mar 2020
World Health Organization (2020b) Mental health and COVID-19. http://www.euro.who.int/en/health-topics/health-emergencies/coronavirus-covid-19/novel-coronavirus-2019-ncov-technical-guidance/coronavirus-disease-covid-19-outbreak-technical-guidance-europe/mental-health-and-covid-19. Cited 30 Mar 2020
Xiao Y, Torok ME (2020) Taking the right measures to control COVID-19. Lancet Infect Dis. https://doi.org/10.1016/S1473-3099(20)30152-3
Yang Y, Li W, Zhang Q, Zhang L, Cheung T, Xiang Y-T (2020a) Mental health services for older adults in China during the COVID-19 outbreak. Lancet Psychiatry 7(4):e19
Yang P, Liu P, Li D, Zhao D (2020b) Corona Virus Disease 2019, a growing threat to children? J Infect. https://doi.org/10.1016/j.jinf.2020.02.024
Yi Y, Lagniton PN, Ye S, Li E, Xu R-H, Zhong B-L et al (2020) COVID-19: what has been learned and to be learned about the novel coronavirus disease. Int J Biol Sci 16(10):1753–1766

Correction to: Transmission Cycle of SARS-CoV and SARS-CoV-2

Tushar Yadav and Shailendra K. Saxena

Correction to: Chapter 4 in: S. K. Saxena (ed.), *Coronavirus Disease 2019 (COVID-19)*, Medical Virology: from Pathogenesis to Disease Control, https://doi.org/10.1007/978-981-15-4814-7_4

The published version in Figure 4.1 has been withdrawn in Chapter 4 due to absence of permission to reuse. The corrected figure includes a "redrawn" image.

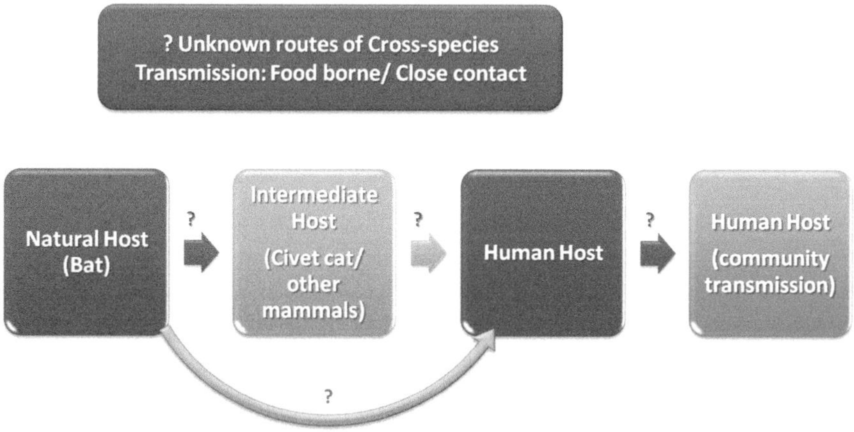

Fig. 4.1 The probable transmission path of SARS-CoV and SARS-CoV-2 from natural hosts to various hosts

---

The updated online version of this chapter can be found at
https://doi.org/10.1007/978-981-15-4814-7_4

S. K. Saxena (ed.), *Coronavirus Disease 2019 (COVID-19)*, Medical Virology: from Pathogenesis to Disease Control, https://doi.org/10.1007/978-981-15-4814-7_17

Correction to: Transmission Cycle of SARS-CoV and SARS-CoV-2